safety
sharingtheexperience
improving the way lessons are learned
through people, process and technology

BP Process Safety Series

Liquid Hydrocarbon Storage Tank Fires: Prevention and Response

A collection of booklets describing hazards and how to manage them

This booklet is intended as a safety supplement to operator training courses, operating manuals, and operating procedures. It is provided to help the reader better understand the 'why' of safe operating practices and procedures in our plants. Important engineering design features are included. However, technical advances and other changes made after its publication, while generally not affecting principles, could affect some suggestions made herein. The reader is encouraged to examine such advances and changes when selecting and implementing practices and procedures at his/her facility.

While the information in this booklet is intended to increase the store-house of knowledge in safe operations, it is important for the reader to recognize that this material is generic in nature, that it is not unit specific, and, accordingly, that its contents may not be subject to literal application. Instead, as noted above, it is supplemental information for use in already established training programmes; and it should not be treated as a substitute for otherwise applicable operator training courses, operating manuals or operating procedures. The advice in this booklet is a matter of opinion only and should not be construed as a representation or statement of any kind as to the effect of following such advice and no responsibility for the use of it can be assumed by BP.

This disclaimer shall have effect only to the extent permitted by any applicable law.

Queries and suggestions regarding the technical content of this booklet should be addressed to Frédéric Gil, BP, Chertsey Road, Sunbury on Thames, TW16 7LN, UK. E-mail: gilf@bp.com

All rights reserved. No part of this publication may be reproduced, stored in a retrieval system, or transmitted, in any form or by any means, electronic, mechanical, photocopying, recording or otherwise without the prior permission of the publisher.

Published by
Institution of Chemical Engineers (IChemE)
Davis Building
165–189 Railway Terrace
Rugby, CV21 3HQ, UK

IChemE is a Registered Charity in England and Wales
Offices in Rugby (UK), London (UK), Melbourne (Australia) and
Kuala Lumpur (Malaysia)

© 2008 BP International Limited

ISBN-13: 978 0 85295 528 4

First edition 2003; Second edition 2005; Third edition 2006; Fourth edition 2008

Typeset by Techset Composition Limited, Salisbury, UK
Printed by Henry Ling, Dorchester, UK

It should not be necessary for each generation to rediscover principles of process safety which the generation before discovered. We must learn from the experience of others rather than learn the hard way. We must pass on to the next generation a record of what we have learned.

Foreword

Hydrocarbons tank fires have proved to be very demanding emergencies due to their size and duration. This booklet is intended for those operators, engineers and firefighters working on, or near, tank farms so that they are aware of the potential emergencies and their consequences, and can adopt safe designs and practices to minimize the occurrence of such incidents and identify the correct responses.

I strongly recommend you take the time to read this book carefully. The usefulness of this booklet is not limited to operating and firefighting people; there are many useful applications for the maintenance, design and construction of facilities and for emergency response.

Please feel free to share your experience with others since this is one of the most effective means of communicating lessons learned and avoiding safety incidents in the future.

Greg Coleman, Group Vice President, HSSE

Acknowledgements

The co-operation of the following in providing data and illustrations for this edition is gratefully acknowledged:

- BP Group Fire Advisor and BP Refining Fire Engineering Advisor
- Dr Niall Ramsden, Paul Watkins and John Frame of Resource Protection International
- Paul Fox of Photo N°1 Fire Services, Hawaii
- Energy Institute (UK) and API
- Ken Palmer, IChemE Loss Prevention Panel member
- Williams Fire and Hazard Control
- René Dosne

Contents

1	**Introduction**	1
2	**Tank design**	2
3	**Initiating events**	4
3.1	Tank fire scenarios	7
3.2	Ignition sources	8
3.3	Static from foam and sunken roof management	12
4	**Fire prevention**	14
5	**Maximum feasible extinguishment**	16
6	**Foam firefighting**	19
6.1	Foam application	19
6.2	Firefighting equipment	29
7	**Firefighting techniques**	31
7.1	Full surface fires	31
7.2	Rimseal fires	39
7.3	Bund (dike) fires	41
7.4	Foam supplies	43
7.5	Water supplies	44
8	**Conclusions**	46

Appendices:

1.	Short bibliography	47
2.	Critical application	50
3.	Escalation	51
4.	Pre-fire plan checklist	53
5.	Specific hazards	55
6.	Properties of foams and other extinguishants	61
7.	Firefighting equipment	71
8.	Some critical questions	91
9.	Learning from past accidents	92

Note: all units in this booklet are metric and US (a US gallon equals 0.83 UK gallon).

1
Introduction

Fires in petroleum product storage tanks are, fortunately, rare occurrences. However, when they do occur they require considerable resources both in manpower and equipment in order to extinguish successfully. Some of the causes of tank fires are outlined in chapter 3. In view of the low number of tank fires on record, relatively few people have had direct experience with fighting tank fires. This document has been prepared to help remedy this deficiency.

This booklet should be used as a training document only. For more in-depth guidance, the API 2021 fourth edition of May 2001, current NFPA Standard 11, BP Guidance Note n°17 on 'Oil tank fires' and other documents listed in the bibliography should be consulted.

It is also important to remember that once started, even if they look impressive, tank fires are not usually a life threatening hazard, as long as good practice is applied.

A major study, known as the LASTFIRE Project, has been carried out by 16 oil companies to review the risks associated with fires in open top floating roof storage tanks. This has now become the definitive study into this subject and many of its findings have been incorporated into this document. This booklet wholly endorses the findings of the LASTFIRE study and the subsequent work carried out on foam testing.

Rimseal fire at an early stage

Note: To complement this book, BP Refining Fire Community of Practice produced two double slide rules to use for training purposes. See Section 7.5 for more details.

2
Tank design

There are three main different types of tank for storing liquid hydrocarbons in large quantities:

- fixed (also called 'cone') roof tanks;
- fixed roof tanks with internal floating roof (also called 'floating screen');
- open top floating roof tanks (simple pontoon or double deck).

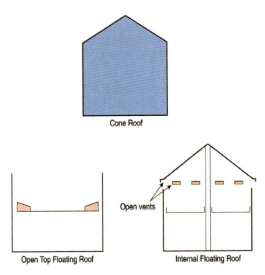

As a general rule, fixed roof tanks are used for 'black', heavy products (heavier than jet/kerosene/gasoil/diesel/naphtha) such as fuel-oils, atmospheric or vacuum residue and asphalt (bitumen). Therefore, they are often fitted with accessories such as steam or oil coil heating and insulation.

Open top and internal floating roof tanks are mainly dedicated to products capable of emitting large quantities of vapours at ambient conditions such as:

- crude oil;
- 'white' light products like jet, diesel, gasoline.

As their roof is floating directly on top of liquid, this design prevents the formation of a flammable mixture of air/hydrocarbon vapours which would occur in a fixed roof tank.

Internal floaters increase protection from fire exposure (therefore fitting a geodesic dome over a floating roof tank significantly decreases the probability of ignition).

More details are given in Chapter 5 and in the BP Process Safety Booklet *Safe Tank Farms and (Un)loading Operations*.

For fire prevention and firefighting purposes, it is important to note that tanks may be fitted with a very wide range of accessories (mixers equipments, inerting systems, instrumentation monitoring (level, temperature...), controllers, fire proofed valves...) and that each site should maintain an up-to-date database of its tanks, their specifications and the product they routinely contain. Also, it is important to know where the product comes from and how process upsets/deviations can modify it. The next two accidents are illustrations of why this is essential:

ACCIDENT The figure below shows an incident which occurred when a 15 bar steam heating system was mistakenly left on for several days, on an atmospheric residue tank containing water (as is often the case with product received from ships). When the temperature was enough to vaporize the trapped water, this happened instantly and damaged the tank beyond repair. Hot product was also projected over a large area. This could have resulted in a serious fire, had an ignition source been found.

ACCIDENT

An explosion and a fire occurred when lightning struck this fuel-oil tank. The investigation showed that the fuel-oil contained enough propane to create a flammable atmosphere below the roof (fuel-oil stream from propane deasphalting unit).

3
Initiating events

The LASTFIRE study listed the most common initiating events for large tank fires.

For fixed/cone roof tanks:

1. unexpected flammable/explosive mixture in the tank;
2. flammable/explosive mixture in normal operation;
3. overpressure;
4. high temperatures/autoignition;
5. holes in roof;
6. overfilling;
7. leakage from tank bottom or shell;
8. leakage/spillage in bund during preparation for maintenance;
9. external event (terrorism, earthquake, flare, escalation from another tank . . .).

For floating roof tanks:

1. failure of pontoon or double deck roof;
2. accumulation of liquid on the roof;
3. tank overfilled;
4. ignition by lightning of flammable vapour in rim seal area;
5. leakage from tank bottom or shell;
6. leakage from side-entry mixers;
7. backflow of liquid onto the roof from the emergency drain on pontoon roofs;
8. leakage/spillage in bund during preparation for maintenance;
9. external event (terrorism, earthquake, flare, escalation from another tank . . .).

We can also add:

- misapplication of foam generating static spark (see end of this chapter);
- ignition by pyrophoric scale deposits;
- ignition by non explosion-proof electric equipment;
- hot work;

- introduction of a product with too high True Vapour Pressure (TVP) (such as injecting too much butane in a gasoline tank).

See examples in this booklet and in the BP Process Safety Booklet *Safe Tank Farm and (Un)loading Operations* (ISBN 978 0 85295 509 3).

The following two graphs are extracted from the LASTFIRE study for large floating roof tanks:

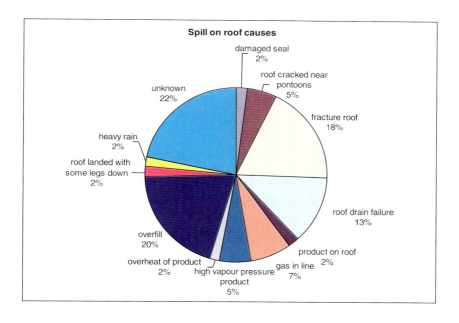

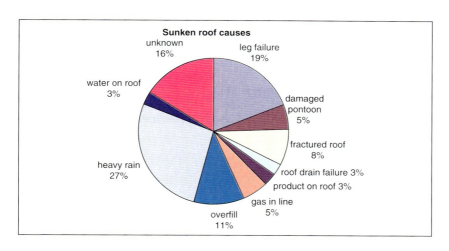

> **ACCIDENT** An accident occurred when the roof of a Jet A1 tank began to be covered with product after heavy rain. The tank had just been put back in service after routine inspection and repairs.
>
> These repairs included changing some metal sheets of the simple deck roof. The investigation revealed that during that job, the contractor replaced the emergency drain pipe (which is supposed to send rain water into the tank in case the normal drain is closed or plugged, to prevent overloading of the roof) by a longer pipe than the original one. Therefore, more rain was allowed to stay on the roof, and the weight from the roof and rain forced Jet A1 to flow back through the emergency drain and flooded the roof.
>
> *Normal design of the emergency drain—in case of heavy rain, water is allowed into the tank to prevent the roof from sinking.*
>
> *Emergency drain modified—weight of water pushes the roof down, forcing product on top of the roof.*
>
> Note that emergency drains on single deck floating roofs are not recommended because of the risk of product backflow onto the roof. These tanks should be equipped with a sufficient number and size of normal rainwater drains with outlet valves kept opened, and should be regularly inspected to ensure their continued integrity.

However small and cheap a modification appears to be, it must still be subjected to your site's 'Management of Change' procedure to ensure all potential hazards have been adequately addressed.

3.1 Tank fire scenarios

The method of dealing with a tank fire will depend upon the type of construction of the tank roof.

An explosion *in a fixed roof tank* will generally result in the weak tank shell to roof joint opening for only part of the tank circumference. In tanks of small or medium diameter the complete roof may be lost (see Appendix 5). The effect of only being able to apply foam through this 'fishmouth' (as illustrated by the picture below) can mean that it may be necessary to attempt to tackle the fire from inside the bunded (diked) area with its inherent risks to firefighting personnel.

Internal floating roof tanks should be tackled in the same manner as fixed roof tanks, as the internal roof is of light construction and will rapidly break up under the effects of the fire.

Fires in *floating roof tanks* can either be:

- in the seal area;
- on the roof itself due to the presence of product;
- full surface because a seal fire or fire on the roof was not dealt with promptly, or because the roof has sunk, either prior to the fire or as a result of poor firefighting techniques. Particularly difficult to extinguish are those fires where the roof is partially submerged as it will be difficult for the foam to flow under the overhanging angled roof.

> The LASTFIRE study showed that rimseal fires are the most common scenario. They are unlikely to escalate to full surface fires in well maintained tanks (some rimseal fires have been known to last for weeks without escalation).

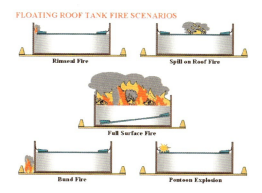

3.2 Ignition sources

Lightning is the most common ignition source. Correlations between rimseal fire frequency and thunderstorm frequency have been developed in the LASTFIRE study. Typical frequency for Northern Europe sites is 1×10^{-3}/tank year; 2×10^{-3}/tank year for Southern Europe, North America and Singapore; and up to 13×10^{-3}/tank year in Venezuela or Thailand; and 21×10^{-3}/tank year in Nigeria.

Therefore, a refinery having 50 large floating roof tanks in the US or Southern Europe statistically has one rimseal fire every 10 years (with possible escalation) ($50 \times 2 \times 10^{-3} = 0.1$ fire/year => 1 fire/10 years).

Picture of a floating roof to shell shunt test (submitted to a 830 A current to simulate lightning) showing the sparks generation (note that wax and rust deposits increase sparking).

It is very likely that such sparking will ignite any vapour present near the seal area, emphasizing the importance of seal integrity.

Picture from tests by Culham Electromagnetics and Lightning Limited for the Energy Institute (UK) and API.

However, other sources are not uncommon, such as:

- operators investigating a suspected leak with an engine driven vehicle;
- hot work;
- pyrophoric deposits;
- static electricity;
- plant flare;
- outside activity (for example, waste disposal field sending burning cardboard on top of floating roof tanks...), etc.

ACCIDENT

This is what is left of the car of operators rushing to investigate a suspected gasoline leak.

Operators were killed and the fire lasted for days, destroying numerous tanks.

ACCIDENT A vacuum bottoms tank's shell to roof weld joint failed spilling hot oil in the surrounding dike/bund. This resulted in a dike/bund fire which was eventually extinguished after approximately two hours. Investigators considered that the most probable cause of the weld failure was a minor internal explosion/overpressure due to the ignition of flammable vapour by pyrophoric deposits.

ACCIDENT Another accident occurred when a Fluid Catalytic Cracker Unit was started after a turnaround. Liquid was sent to the flare and ignited a water treatment tank (without roof) 140 m (460 ft) away. The tank contained water contaminated with the crude from the crude unit desalter. Are water treatment tanks included in your emergency response prefire plans? Do you have enough hydrants nearby?

ACCIDENT An explosion occurred in a waterflood header tank. It was ignited by welding repairs to an inlet nozzle. Unknown to the three contractors working on the tank, there was an explosive gas mixture inside the tank. All three employees received bruising and abrasions from the incident.

For hot work, it is important to note that product can be trapped in many places, as the figure below illustrates.

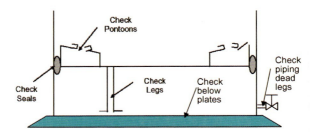

ACCIDENT A 5000 barrel crude tank was being cleaned when an explosion lifted it several feet off the ground, splitting the roof open 1/3 to 1/2 the circumference at the roof seam and shooting a yellow flame horizontally 20 to 30 feet (6–9 m) out of the roof opening. The vapours coming from an open hatch ignited on the 300 v DC line that was left seven years before when an ultrasonic level sensor was dismantled.

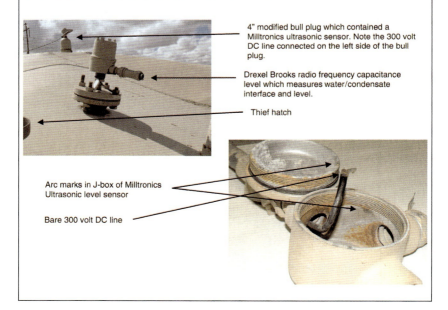

ACCIDENT Batch blending was going on in a 7,000 m³ unleaded gasoline tank when a fire occurred. During more than 30 hours, 56 fire trucks tried to tackle the fire, successfully protecting adjacent tanks. The investigation team found that the blending calculations were wrong: three times too much butane was being sent to the tank. A bubble of light ends probably lifted and tilted the roof, creating enough static or metal to metal friction to ignite the vapours.

Bad control of blending operations has caused multiple floating roof sinkings (see the Booklet *Safe Handling of Light Ends* (ISBN 978 0 85295 478 2) in this series).

Refer to the booklet *Safe Tank Farm and (Un)loading Operations* (ISBN 978 0 85295 509 3) in this series for more information on different tank incidents.

3.3 Static from foam and sunken roof management

It has become apparent that a number of tank fires which hitherto have been recorded as 'cause unknown' have been caused by static electricity generated during the application of foam from firemen's nozzles or remote monitors. Indeed, the re-ignition of fires may be related to foam application.

Refer to appendix A9.4 for an example of such an incident.

ACCIDENT An accident was caused by foam application on exposed naphtha after the floating roof of a storage tank sank. Static created by foam application ignited the fire that it was supposed to prevent. As a result of escalation, three naphtha tanks were destroyed as the figures below illustrate.

a. Roof sunk

b. Tank ignited by foam application

c. Adjacent tank beginning to burn

d. Two tanks fully involved

e. Three tanks fully involved

Sunken roof management

In the case of a large exposed surface of refined product (such as a sunken roof on a jet tank):

1. *Stop all transfer of product on or out of the tank:*
   - Assess the situation and determine the hazardous area using gas testers.
   - Make sure that there is no close ignition source and evacuate personnel.
2. *DO NOT USE FOAM*, except:
   - if there is a higher probability of ignition by a non-removable ignition source (such as a lightning storm);
   - if personnel must be protected against fire during the subsequent operations (for example, removal of product, roof repairs);
   - if the product involved has a high conductivity (such as crude oil).
3. *IF A DECISION IS MADE TO APPLY FOAM*:
   - If possible, use fixed pourers so as to apply foam as gently as possible down the tank shell.
   - Foam generated by monitors or hand held nozzles should be applied on the internal shell of the tank before going on the product.
   - Fire appliances with integrated foam proportionners are preferred to portable foam proportionners.
   - If portable foam proportionners are used, the maximum foam flow must first be generated outside the tank and then applied as gently as possible on the internal shell of the tank before going on the product.
   - Never apply foam or water directly to the surface of the hydrocarbon product.
4. *If a foam cover was established on a refined product (after a fire or after conditions of the above chapter):*
   - Once the foam cover is created, maintain it regularly and gently.
   - Keep a close watch on the tank until all product is removed.
   - The natural degradation of the foam cover may lead to an electrostatic ignition by the foam and water sinking through the hydrocarbon product.

ACCIDENT

The roof of this tank was damaged during an earthquake. Foam was applied as a preventive measure using foam pourers but the foam blanket was not maintained. Ignition occurred because of static build-up where a foam pourer maintained a continuous dripping of water and foam onto the naphtha surface.

4

Fire prevention

The LASTFIRE study showed that many tank major incidents were due to simple practices being forgotten or overlooked. *The most efficient technique to prevent tank fires or major leaks is to adhere to the good practices briefly mentioned below (refer to the BP Process Safety Booklet Safe Tank Farms and (Un)loading Operations for more details).*

Operations

- monitor tank fill/discharge levels as a routine;
- respond to high level or low level alarms;
- react to any level alarms (even if trips are provided);
- if high-high alarm: visual check of tank (overfilling into bund);
- prevent roof «landing» => air entry or damage;
- hazop routine and non-routine operations;
- safeguard against product transfer errors (high RVP, hot product...);
- have clear and up-to-date emergency operation procedures and train operators.

Monthly formal checks by operator

- cleanliness of roof;
- leakage signs;
- roof drains (including emergency one);
- pressure valve vent mesh;
- weather shields/seals;
- pontoon compartments (water, oil, LEL test, covers tight...);
- earthing cables;
- guide poles;
- rolling ladder;
- roof drain valves;
- bottom of shell.

Example of a roof with waxy deposits

Examples of roofs showing leaks of product

ACCIDENT This rimseal fire escalated quickly to a full surface fire when vapours contained in leaking pontoons exploded. While the rimseal fire might have been dealt with, the full surface fire proved difficult to extinguish because of a lack of water resources. The site had no routine practice of gas-testing pontoons regularly and therefore, escalation was inevitable.

Fire once the roof lost buoyancy

Early stage of the rimseal fire as captured by security camera. Note flying pontoon plate in yellow circle

Example of poor design of a pontoon manhole cover. There is no gas-testing hatch. (Also note that the foam dam is lower than the secondary seal, which was fitted later to the tank for environmental reasons—this denotes poor Management of Change and poor understanding of foam systems).

Manhole covers should be secured as loose covers can float away when the roof starts to sink, or they can be blown away by wind or fire water streams. A gas-testing hatch should be provided for each compartment.

5

Maximum feasible extinguishment

According to evidence from actual incidents that have occurred throughout the world over the past few years, extinguishing a full surface fire in a large tank (over 46 m (150 ft) in diameter) using mobile equipment is feasible (tanks up to 83 m (272 ft) in diameter have been successfully extinguished using only mobile equipment) but needs careful planning, large delivery devices and support equipment and well trained teams of operators. In accordance with the LASTFIRE study, a risk analysis should be carried out to assess the feasibility and justification of attempting to extinguish full surface fires. In the event that it is thought feasible then mobile monitors or fixed systems can be considered according to local circumstances.

Reputation and media attention issues should be included in any assessment.

This full surface 83 m (270 ft) diameter tank fire was successfully extinguished using mobile equipment (application rate roughly 8.8 l/m²/min (0.21 gpm/ft²), no wind or rain)

Typical scenarios that must be included in these formal risk assessments are, considering a tank farm only (other issues such as pump rooms fire, loading gantries fire should also be considered as part of the emergency planning but are outside the scope of this booklet):

- three dimensional fire at tank bottom;
- rimseal fire for floating roof tanks;
- vent fire for fixed roof tanks;

- full surface tank fire;
- bund spill fire;
- full bunded area fire.

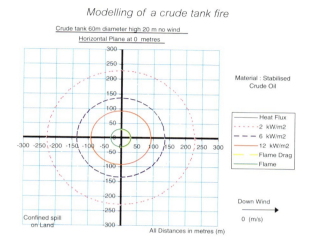

Modelling of a crude tank fire

These formal risk assessments should use models such as BP CIRRUS to evaluate the consequences of each scenario (for example, thermal radiation). A fire is very unlikely to escalate to adjacent tanks if the radiation levels on the exposed tank are kept below a level of 8 kW/m². Fire modelling can be used to assess, under typical site environmental conditions, how far apart tanks need to be to achieve this.

Typical outcome of the QRA is a choice between:

- fight this scenario with fixed or semi-fixed equipment;
- fight this scenario with mobile equipment;
- pump the product out, let the tank burn and cool exposed adjacent tanks.

This last option is perfectly acceptable in some situations, for example, remote storage in desert with limited water supply (see appendix 5). The following plans therefore need to be prepared for such a contingency:

- how and to where the product in the affected tank will be pumped;
- the necessity for protecting adjacent tanks;
- the effect of boilover or slopover (overfill).

It is important to include a fire specialist to consider life threatening hazards while studying the possible strategies:

- Except in very rare occasions, products susceptible to boilover should not be left to burn out (see Appendix 5).
- Generally speaking, the policy is that *no person should have to go onto the roof of a floating roof tank to extinguish a rimseal fire*. However, in some cases, it might be (and has been) the only option, particularly if there is no safe walkway around the wind girder and there is no fixed foam system. In this case, it should be done to a preplanned response which includes completion of a Job Safety Analysis included in the emergency plan (further information on safety aspects of this can be found in the LASTFIRE video and documents). Firefighters are warned not to get onto floating roofs during a rimseal fire unless there is a good floating roof pontoon inspection program in place and the chance of pontoon explosions due to heat are limited to a very low probability.
- Firefighting strategies should not normally require firefighters to enter a bund to install monitors when a tank is on fire in that bund (see monitor range considerations in appendix 7), although sometimes this might be the only option.

Example of a damaged pontoon after an internal explosion. This is why every means of fighting rimseal fires from the wind girder with portable equipment or from the ground, via fixed or semi-fixed systems, should be in place. Firefighting from the ground level using monitors should not be used due to the possibility of tilting the roof.

Portable equipment has been specifically designed to be manually attached to the shell of a tank on fire (see the following pictures and refer to Section 7.2) and has been used successfully in June 2003 on a rimseal fire on a BP site.

6

Foam firefighting

6.1 Foam application

With few exceptions, extinguishing a fire in a petroleum storage tank will require the application of a foam concentrate/water solution at a rate sufficient to be able to cause a blanket of aerated foam to cover the surface of the burning liquid, thus eliminating the air.

Failure to achieve this 'critical application rate' (Appendix 2) will see the foam blanket destroyed at a faster rate than it can be produced and therefore the fire will not be extinguished.

Manufacturers of foam concentrates and the NFPA National Fire Codes give recommended application rates for use with particular products which depend upon the method of application. *These recommended rates are based upon the assumption that all the foam will reach the surface of the burning liquid.* The foam concentrate must be of good quality and maintained in good condition by proper storage and testing. The use of the LASTFIRE fire test is recommended for evaluating foam for storage tank application.

Using portable equipment

Liquid hydrocarbons (with no more than 15% alcohol by volume*):

It is recommended that when using portable foam monitors to apply foam, *the rate that foam is produced at grade level should be increased by up to 60% over recommended minimum NFPA rates* to allow for the loss of foam which

Foam application through a portable monitor.

* This includes gasoils and motor spirits containing no more than 15% alcohol (MTBE or ETBE) by volume. Once this percentage is exceeded, the product should be considered as a water soluble fuel and the concentrate should be used as is recommended by the manufacturer for such a fuel (3 to 6%).

fails to reach the tank interior and breaks down due to heat and thermal currents (the latter have been recorded upwards of 80 km/hr (50 mph)), inexpert operation of monitors and variations in wind speed/direction.

Foam losses can be caused by a number of combined causes such as insufficient range, high wind, foam/monitor quality, tank deformed...

The guidance given in NFPA, strictly speaking for tanks up to 18m (60 ft) diameter, is that if monitor attack is to be used for a full surface fire in crude oil and light product tanks, the foam solution should be applied at a rate sufficient to ensure an applied rate at the surface of the liquid of 6.5 l/min/m^2 (0.16 gpm/ft^2). In order to ensure this, it will be necessary to generate 10.4 l/min/m^2 (0.26 gpm/ft^2) i.e. 6.5 l/min/m^2 (0.16 gpm/ft^2) plus 60%.

Flammable liquids having a boiling point of less than 100°F (37.8°C) may require higher rates of application. Flammable liquids with a wide boiling range may develop a heat layer after prolonged burning and can require application rates of 8.1 l/min/m^2 (0.2 gpm/ft^2) or more (therefore recommended rate 12.9 l/min/m^2 (0.32 gpm/ft^2)). (See also Appendix 5)

Other flammable/combustible liquids

Water soluble, certain flammable and combustible liquids and polar solvents are destructive to regular foams and require the use of alcohol resistant foams. In most instances a 6% foam solution will be necessary, however some suppliers now provide 3% alcohol resistant foams.

Liquid	NFPA application rate	Recommended application rate (NFPA +60%)
Methyl alcohol–Ethyl alcohol–Acrylonitrile–Ethyl acetate–Methyl ethyl ketone	6.5 l/min/m^2 (0.16 gpm/ft^2)	10.4 l/m^2/min (0.26 gpm/ft^2)
Acetone–Butyl alcohol–Isopropyl ether	9.8 l/min/m^2 (0.25 gpm/ft^2)	15.7 l/m^2/min (0.4 gpm/ft^2)

Discharge duration

Products with a flash point between 100°F (38°C) and 200°F (90°C) (kerosene)	50 min.
Products with a flash point below 100°F (38°C) (gasoline)	65 min.
Crude oil	65 min.

Foam monitors should all be sited at the one location with the foam streams entering the tank at the same point (see figure below) and impinging on the surface in the same area. This will help establish a stable foam blanket quicker and more effectively than applying the foam on the surface at three or four separate locations. It is application density that establishes a *bridge head*.

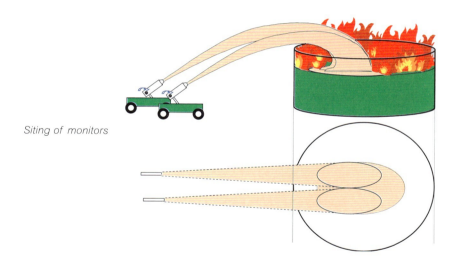

Siting of monitors

The sooner a large pool of foam is established (or a foot print as it is sometimes called), the sooner the fire will be put out.

Externally cooling the tank shell in the region of the liquid level may assist the foam in sealing against the hot tank walls, but cooling water streams should only be played onto the shell once the foam blanket has achieved good control of the fire. Care has to be exercised when using any water streams during foam application since they may dilute the foam blanket being formed if the streams break up and water 'drifts' into the foam. Also, application of water cooling may distort the tank shell.

Water applied to the shell of a burning tank is normally ineffective and a waste of resource. However, such cooling may assist in the late stages of extinguishment of a full surface fire or rimseal fire, as the cooling of the area of the liquid level allows the foam to seal against the hot tank wall. Water should be reserved for the immediate protection of exposures being subjected to radiated heat.

Using fixed equipment

The best protection for storage tanks containing flammable liquid is the provision of fixed firefighting equipment. The use of portable foam equipment to extinguish a full surface fire is difficult and fraught with danger. There are numerous documented cases where failure to extinguish a fire can be directly linked to the absence of a fixed protection system.

Lower application rates than with mobile equipment are permissible when using fixed fire equipment such as subsurface (base injection) systems or fixed foam pourers (see figures on pages 22 and 23).

Types of systems

There are three main types of systems currently in use designed to enable the aerated foam concentrate/water mixture to reach the surface of the burning liquid:

1. **Subsurface foam injection system**. Designed to discharge foam into the base of a tank either through product lines or separate specific pipework. The foam floats to the surface of the liquid and is not affected by the flames or thermal updraft. Not suitable for floating roof tanks, for cone tanks with internal floating roofs or water soluble fuels. There are also semi–subsurface* injection systems that are designed to protect the foam from the hydrocarbons, but these are strongly not recommended due to the difficulty of testing them.

2. **Rimseal foam pourer system**. Designed to place foam in the area of the rimseal of a floating roof tank. A foam/water solution is injected into a pipework system from outside the bund area, and it is aerated and allowed to fall into a dam constructed around the seal. There are a number of variations of this system.

3. **Top foam pourer system**. Designed to place foam on the surface of a liquid through pipework accessed from outside the bund. Can be used on fixed (cone) roof, floating roof and internal floating roof tanks.

---

*The equipment used for the semi-subsurface technique consists of a container, either mounted in the fuel itself or just outside the tank shell near its base, with a hose having a length greater than the height of the tank. The non-porous foam discharge hose is made from a synthetic elastomer coated nylon fabric and is lightweight, flexible and oil resistant. It is packed into the container in such a way that it can easily be pushed out by foam entering it from a foam generator. The container is provided with a cap or bursting disc to exclude products from the hose container and foam supply piping.

Subsurface injection (roof is shown intact for demonstration purposes)

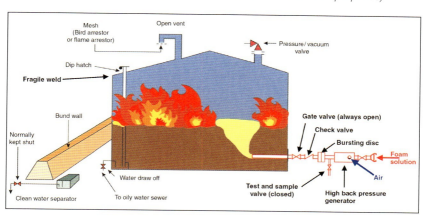

Application rates for subsurface injection

Flash point of contained product*	Solution rate NFPA	Minimum duration (minutes)
flash point > 100°F (38°C)	4.1 l/min/m² (0.1 gpm/ft²)	30
flash point < 100°F (38°C), including crude oil		55

* This includes gasoils and motor spirits containing no more than 15% alcohol (MTBE or ETBE) by volume. Flammable liquids having a boiling point of less than 100°F (37.8°C) may require higher rates of application. Water soluble and certain flammable and combustible liquids and polar solvents are destructive to regular foams and require the use of alcohol resistant foams.

Preference is for top pourers rather than subsurface system as the latter are more difficult to maintain and cannot be easily tested unless special test points are incorporated into the design (see also Appendix 6). They do not allow change of product in the tank (from non-foam-destructive to foam destructive) or upgrading of the tank (from simple fixed roof to fixed roof with internal cover float). These systems also represent a potential liquid leak source. However, it must also be recognized that top pourers are more vulnerable to damage from an internal explosion or the subsequent fire.

Subsurface application is ineffective on polar solvents because the foam dissolves and topside application is required.

LIQUID HYDROCARBON STORAGE TANK FIRES

Top pourer foam application (roof is shown intact for demonstration purposes)

Note: Some bursting discs are vertical along the shell tank, which is a better option than the one shown below (gas free atmosphere when opening foam pourer).

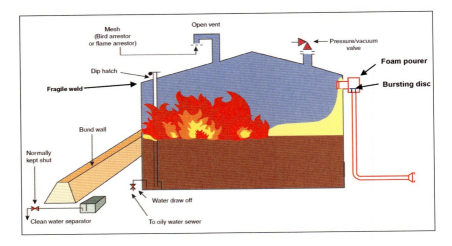

Application rates for top pourers for fixed roof tanks

Flash point of contained product*	Solution rate	Minimum duration (minutes)
Tanks below 45 m (150 ft) diameter (NFPA)		
flash point > 100°F (38°C)	4.1 l/min/m² (0.1 gpm/ft²)	20** 30***
flash point < 100°F (38°C), including crude oil		30** 55***
Tanks above 45 m (150 ft) diameter (BP)		
All products	6 l/min/m² (0.15 gpm/ft²)	30** 55***

* is as * on page 22 for subsurface systems.
** a discharge outlet that will conduct and deliver foam gently onto the liquid surface without submergence of the foam or agitation of the surface.
*** a discharge outlet that does not deliver foam gently onto the liquid surface but is designed to lessen submergence of the foam or agitation of the surface.

LIQUID HYDROCARBON STORAGE TANK FIRES

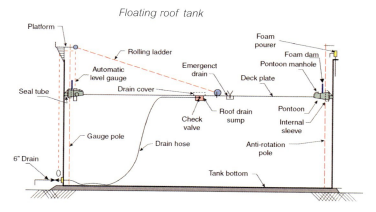

Floating roof tank

Application rates for top pourers for floating roof tanks

Type of system*	Solution rate NFPA	Minimum duration (minutes)
Rimseal pourer (with foam dam)	12.2 l/min/m² (0.3 gpm/ft²)	20
Rimseal pourer (without foam dam)	20.4 l/min/m² (0.5 gpm/ft²)	10
Full surface fire pourers	See table for fixed roof tanks	

* is as * on page 22 for subsurface systems.

For a fire in the roof seal of a floating roof tank, foam solution should be applied at the rate of 12.2 l/min/m² (0.3 gpm/ft²) when a foam dam is fitted, or at the rate of 24.2 l/min/m² (0.6 gpm/ft²) in the absence of a foam dam, based upon a nominal dam width of 600 mm as specified in the NFPA Codes (Appendix 5).

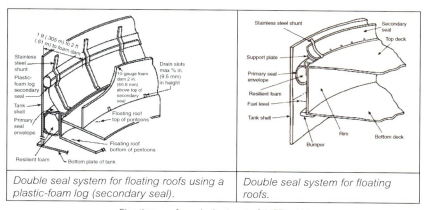

Double seal system for floating roofs using a plastic-foam log (secondary seal).	*Double seal system for floating roofs.*

Floating roof seals (extract of NFPA 11)

25

Integral foam dams (dam located directly above edge of the pontoon, closer to the shell than traditional NFPA11 dams see picture below) should be preferred as they prevent water and oil accumulation on the pontoons and allow fast accumulation of foam while reducing the amount of foam necessary.

However, dams designed in accordance with NFPA11 are acceptable. For NFPA11 dams, drain slots are mandatory and shall be big enough to allow rain water/foam water to flow, including an allowance for debris accumulation; but they should not be too big to prevent foam build-up (see right hand side picture above).

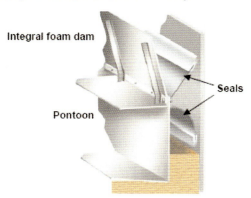

Rim seal fires

ACCIDENT A 500,000 bbls (80,000 m^3) open top floating roof tank 91% full with crude oil was struck by lightning. The resulting rimseal fire (60% of the circumference) and small roof spot fires were completely extinguished within 90 minutes of ignition.

The tank was used for high pour point crudes which left a wax residue on the inside of the shell because of ineffective seal scrapers. When the wax melted in the sun, it ran onto the roof and plugged the internal drain. The wax also acted as insulation preventing a good contact for the metallic shunts and an increased potential gap for sparks. The tank was equipped with primary and secondary seals.

The tank was 75 m (246 ft) in diameter and fitted with a rim foam distributor system (central foam distribution manifold located on the roof—see Appendix 7) and had no foam dam.

Application of foam with the fixed system only pushed the fire onto the roof. A team had to climb on the wind girder to tackle the fire using hand-held nozzles and some of the tools shown at the end of Chapter 5. The wind girder was equipped with hand rails which permitted safe access around the tank to fight the spot roof fires using a hand line.

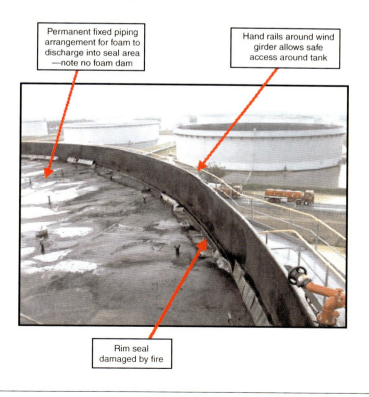

Permanent fixed piping arrangement for foam to discharge into seal area —note no foam dam

Hand rails around wind girder allows safe access around tank

Rim seal damaged by fire

Foam concentrate

The preferred type of foam for application on a petroleum storage tank fire should be a concentrate that provides good burnback resistance and also rapid knockdown characteristics—foam should have been selected as part of the LASTFIRE tests. The correct expansion ratio and flowability of foam concentrate are two further critical factors to be considered. Appendix 7 gives guidance on these matters.

Many types of foam concentrate are available for use as 1, 3 or 6 percent solutions by volume. Concentrates for use at 3 percent are normally preferred to those for use at 6 percent because of their greater efficiency in use, storage and handling. The equipment used to proportion and distribute the foam must be compatible with the concentrate being used. It is therefore recommended to use 3 or 1 percent concentrations. 1% solution is now available and is considered as efficient as 3% foams (as demonstrated during LASTFIRE tests) but proportioning equipment must be sufficiently accurate at this setting. Also, as 1% foams are less fluid, attention should be given to pumping capabilities, especially in cold weather conditions.

Foam concentrates of different types or from different manufacturers should not be mixed unless it has been established that they are completely compatible.

LASTFIRE fire test of a foam (see Appendix 6 for more details)

6.2 Firefighting equipment

Appendix 7 gives examples and advice on equipment.

The preferred configuration is:

- **Fixed roof <18 m (60 ft) diameter without internal floating deck:** Top pourers for an application rate of 4.1 l/m²/min (0.1 gpm/ft²) over the full surface area of the tank (pourers should be fitted to the shell of the tank not the roof). However use of mobile or portable monitors is acceptable if manpower is available.

- **Fixed roof >18 m (60 ft) diameter:** Top pourers over the full surface area of the tank for an application rate of:
  a. 4.1 litres/m²/minute (0.1gpm/ft²) below 45m (150ft) diameter
  b. 6 l/m²/min (0.15 gpm/ft²) above 45m (150ft) diameter

Pourers should be fitted to the shell of the tank not the roof as the roof is more sensitive to damage from an internal explosion.

- **Fixed roof with internal floating deck:** Top pourers to cover the full surface area.

- **Open top floating roof <35 metres (115ft) diameter:** Top pourers for an application rate of 4.1 l/m²/min (0.1 gpm/ft²) over the full surface area of the tank. (However use of mobile or portable monitors is acceptable if manpower is available). Pourers should be fitted to the shell of the tank not the roof (as they are designed to fight a full surface fire, in which case the roof is sunk).

- **Open top floating roof >35 metres (115ft) diameter:** Fixed or semi-fixed foam system pourers are recommended to fight rimseal fires, the application rate to be achieved with foam dams is 12.2 l/m²/min (0.3 gpm/ft²); without foam dams 20.4 l/m²/min (0.5 gpm/ft²). Where pourers are designed to cover the whole of the tank roof area (if justified by a risk analysis—see Chapter 5), the application rate should be 4.1 l/m²/min (0.1 gpm/ft²). For very large tanks multiple foam supply lines to alternate pourers are recommended to prevent the whole system being disabled through a single system failure. Pourers should be fitted to the shell of the tank not the roof as the roof is not always accessible for maintenance and test operations and the flexible hose needed may be a reliability concern. In addition, should the roof become flooded with either water or product, the roof could sink, taking the foam system with it.

Open top floating roof tank installations can benefit from a foam dry riser terminating at the gauger's platform, together with hand rails installed round the wind girder, to facilitate an attack using portable equipment to extinguish any remaining pockets of fire in a rimseal if fixed/semi-fixed pourers are not totally effective in extinguishing the fire.

An alternative, preferred option is to have foam solution hydrant outlets at the wind girder level connected to the foam system pipework. For large tanks it may be necessary to have several outlets to reach all parts of the circumference with manageable lengths of hose. It is important to ensure that the foam proportioning system can accommodate changes in flow rate when using the hydrants (and also to allow for some blocked pourers).

- **Shell cooling water deluges**: Water cooling of a tank shell is often over-used. Fire modelling should be used to determine the needs for water cooling. Guidance on this subject can be found in documents listed in the bibliography in Appendix 1. If required, water deluges should be sized for a minimum of 2.1 l/m^2/min (0.05 gpm/ft^2) to protect against radiant heat (not direct flame impingement).

Tank shell and roof waterspray tests

It should be noted that localized cooling of a tank on fire (for example, with a single water monitor) will distort the shell—which is why this is not recommended if there is no fixed water deluge.

Typical example of cooling water effect when applied only to one side. The non cooled side is folding and the shell is subjected to extreme stress.

7 Firefighting techniques

7.1 Full surface fires

Satisfactory extinguishment of a petroleum storage tank fire begins with pre-fire planning and therefore much of the following information should have been considered. The person in charge of the fire will have to consider the following points:

Rescue: the need for rescue of injured people.

Life hazard: the potential need for evacuation of personnel (evacuation distances may exceed 600 m (2,000 ft) (see Appendix 5)), based on:

- type of product burning;
- number of tanks burning;
- protection of exposed structures;
- construction of tanks;
- status of tank and tank valves;
- dike/bund fires;
- vent fire;
- seam fire;
- foam supplies;
- water supplies/location;
- siting of foam monitors;
- water drainage.

For more pre-planning guidelines, see bibliography in Appendix 1 and Appendix 4.

Manpower requirements to tackle a major tank fire will, of course, vary depending upon the type, location and nature of the fire, the method of extinguishment required and the availability of trained personnel. The general requirements for any particular tank will need to be determined during the production of the pre-fire plans.

In general it is recommended that firefighting personnel should be given a rest break after approximately three hours of work. This time may need to be reduced depending upon fatigue levels brought about by environmental stress (including heat/cold, breathing apparatus, dehydration . . .). Manpower planning should take account of this aspect.

Early alerting of emergency response teams is essential to afford them the maximum opportunity to extinguish the fire in its incipient stages. To this end it is recommended that consideration be given to the installation of an automatic linear heat detector (LHD) around the rim of a floating roof tank (see the bibliography in Appendix 1 and Appendix 7 for more details).

It is absolutely essential that all personnel involved in frontline firefighting wear full firefighters turnout gear (see bibliography for more details).

Type of product on fire

The product involved will dictate the required foam application rate. This has an immediate impact upon the control of the fire as it determines the logistical support required.

Crude oil and certain heavier oils are prone to the effects of 'boilover' (see Appendix 5). Due consideration will have to be given to the consequences of this for equipment layout, personnel safety and the anticipated time of extinguishment.

Are the quantities of combustion products given off such as to warrant additional safety features? If so what is the need for evacuation? What are the likely requirements for firefighters to use breathing apparatus? What are the hazards after extinguishment (for example, reignition, explosion, toxic vapours if liquid is toxic e.g. benzene).

A methanol fire such as this one will require alcohol resistant (AR) foam.

Note that this type of fire produces very little smoke and sometimes, even the flames are invisible, except with IR imagery.

Note the helicopter dropping water—use of helicopters or aircraft has been tried on multiple storage tank fires and has never added any technical benefit. It can also prove dangerous on a rimseal fire by sinking the roof.

Number of tanks burning

The number of tanks burning will determine the requirements for manpower, equipment, water and the level of exposure protection necessary.

An immediate assessment will be required of the additional support necessary. This should be ordered without delay.

Tanks are too close—escalation likely

Refer to BP Process Safety Booklet *Safe Tank Farms and (Un)loading Operations* for more details on tank farm design and see Appendix 3 of this booklet.

Status of tanks and tank valves

The ullage space in a tank that is on fire can influence the quantity of foam that may be applied. If the tank is at maximum dip there is a possibility of causing an overfill (slopover) due to the amount of water being released by the breakdown of the foam increasing the level in the tank. There is, however, a positive side to tank depth. A full tank means that there is less heat and flame for the foam to travel through and therefore less breakdown due to these factors. In some instances extinguishment has been aided by pumping water into the bottom of a tank to raise the level and enhance the possibility of more foam reaching the surface of the burning liquid. Such a technique should only be used with great care—if water is pumped into a tank on fire there is probably no indication of liquid level height in the tank. Level indicators will almost certainly have been damaged or destroyed, and guessing where the liquid level is could be dangerous and lead to large quantities of burning fuel being spilled into the bund. *This technique should not be attempted in liquids where the risk of a boilover or slopover exists.* A full tank will also assist in dissipating heat away from the tank shell, whereas if the ullage space is considerable there is far more chance of the tank sides folding in. A tank fire in the UK resulted in the sides folding within eight minutes of the occurrence of a full surface fire.

Example of partial fold-in of a tank shell. The liquid level is well visible as the product kept the paint cool and intact.

Final stage complete fold-in

The effect of radiant heat on export transfer pumps situated nearby must also be considered as they may be required for pump out operations. The fire referred to above also required substantial cooling sprays on transfer pumps situated 25 m (80 ft) away.

It will often be necessary for firefighters to operate or cause to be operated a number of the valves found on a tank. The reasons for requiring valve operation will be varied but the following should be taken as a guide:

- Roof drains on floating roofs are normally left in the open position. In the event of a roof sinking it is then possible for product to leak into the bund (dike) area via this valve. It will normally be prudent to close it and, to do so, it will often be the case that firemen have to enter the bund as the valve will be located on the tank shell. Emergency response plans will normally state that the valve shall be closed at the earliest possible opportunity.

- Water drains for draining water from the tank floor of most tanks are located on the tank shell, and it may be decided that, in order to reduce the effects of a boilover, water should be drained from this point. (Draining water from the bottom of a crude oil tank may not necessarily stop a boilover as water can be layered in the crude or caught in a sunken roof). In reality, it is virtually impossible to drain all the water from the bottom of a tank and only a few centimetres are needed to create a big boilover! If a decision is made to open a water drain valve, consideration should be given to the problems associated with closing it at a later point. Also, the outlet could become covered by slow draining water in the bund (dike) and it will not be possible to determine whether water or product is being released.

- Product suction and fill lines which may or may not be fitted for remote actuation are normally close to the tank shell. It may be necessary to use these valves either for removing or putting product into the tank.

Protection of exposed equipment

An immediate assessment should be made of the risk to both the site workforce and the surrounding population. Immediate evacuation, provided it is safe to do so, is often the safest method of protection. Factors to consider are the nature of the product, wind direction, boilover potential and probable time to extinguishment.

All tanks and vessels closer than one tank diameter upwind and two tank diameters downwind may require cooling water applied to their exposed surfaces. Tanks at 90° to the wind direction within one tank diameter may also require cooling. If resources are limited the water should be applied first to tanks containing lighter products; very small tanks and nearly empty tanks. The water should be applied to the side of the tank facing the involved tank and to the roof area of fixed roof tanks. Other exposures, such as pumps and pipelines, should have water applied according to the prevailing circumstances.

Example of cooling adjacent tanks with mobile equipment

Protection of adjoining tanks with water spray should seek to maximize the water contacting the tank shell thus minimizing water run off. The use of excess water on exposures can reduce supply and pressure and overtax drainage facilities. In order to minimize water use, firefighters should aim to supply just enough water to generate steam from hot surfaces. Excessive water run off will flood drainage systems and allow oil spillage to spread, thus increasing the risk of a flash fire from remote ignition sources.

A combination of both the extensive modelling and experiments suggests that a reasonable application rate is 2 l/min/m² (0.05 gpm/ft²) of surface exposed to radiated heat. Older tanks with rough surfaces will need considerably more cooling water than new smooth-skinned tanks.

Perhaps the best practical application of water spray protection, for either fixed systems or mobile systems, is that recommended by NFPA, which suggests that if steam is generated when cooling water is applied, then its application should be continued. If it is not, then the cooling water should be shut off but the test should be repeated at regular intervals.

Despite the application of water spray, **adjacent floating roof tanks** may still be affected by radiated heat causing surface boil-off especially with low boiling point fuels, thus creating a flammable atmosphere at roof level.

Exposed floating roof tanks should receive immediate application of foam to the rimseal area. An early decision is required as to the possibility and advantages of completely covering the roof with foam. This will depend upon the availability of equipment, the proximity of the fire and wind conditions. At the recommended rate of 6.5 l/min/m² (0.16 gpm/ft²) it will take approximately 8 minutes to cover to a depth of 1/2 metre (1.5 ft) and the load on the roof will be around 50 kg per square metre (10.2 pounds/ft²). It should be ensured that roof drains are open and the tank roof is not overloaded. Gentle application techniques should be used to be sure not to tilt the roof.

On a pontoon roof it will only be necessary to cover the area inside the pontoon, i.e. the area of single plate in the centre of the tank. It is not necessary to apply a foam cover to the total roof area if the roof is of the double-skinned type. In all cases advice should be sought from the local engineering department at the pre-fire planning stage. It is preferable to try to cover the roof by using a rimseal foam system (if fitted), whereby foam application continues beyond 20 minutes and foam overflows the foam dam and flows into the roof centre.

Radiant heat may prevent fire crews accessing the wind girder to try foaming the roof via foam handlines. Alternatively, fire crews may need water curtain protection to gain access to the wind girder.

Applying foam from ground-based portable foam monitors in an attempt to foam the roof is not the best option and should only ever be considered as a last resort. Regardless of whichever tactics are reviewed, each should be risk assessed for consequences.

Careful evaluation will be required before the product in adjacent exposed tanks is pumped out, as this could increase the risk i.e. full tanks assist in preventing the temperature of the shell increasing unduly.

Factors influencing escalation are shown in Appendix 3, as are the estimated typical times for hazardous conditions to be generated at an adjacent tank when exposed to a full surface fire in a 50 m (165 ft) diameter tank containing naphtha.

This crude tank fire could not be extinguished due to lack of resources. Despite boilovers occurring, adjacent tanks were unharmed, due to a sound design that included large tank spacing.

Care must be taken to ensure that **nearby LPG and LNG storage vessels** are kept cool at all times. As a general guide they should, if exposed, be kept covered with a film of water either through the fixed spray system or using portable equipment. Any protective water film should be applied to such vessels (if uninsulated/fireproofed) at a rate of 10.2 l/min/m^2 (0.25 gpm/ft^2). As well as the vessel itself, particular attention shall be given to cooling exposed steelwork such as stairways, top bridles and valve platforms, even where LPG and LNG storage vessels are provided with Passive Fire Protection (PFP).

Example of a LPG sphere that required heavy cooling during a nearby tank fire.

Refer to booklets *Safe Handling of Light Ends* (ISBN 978 0 85295 478 2) and *LNG Fire Protection and Emergency Response* (ISBN 978 0 85295 515 4) in this series.

Successfully achieving the last stages of extinguishment can sometimes be very difficult due to folded shell, roof pieces, etc, that prevent foam from reaching in the last pockets of fire (see picture below). Multiple techniques can be used but they must be planned in advance to prevent the foam attack from running out of supplies.

Use of medium expansion foam can give good results in this type of situation as shown in picture below.

7.2 Rimseal fires

On external floating roof tanks

A fire in the seal area can be tackled in a number of ways, the best and most obvious being by utilizing fixed rimseal pourers. However, if these are not available and a manual attack is the only recourse then the following options are available.

- Use water spray protection to enable a crew to ascend to the gaugers platform with hand-held foam equipment. From this point, and from the wind girder if it is safe to do so, an attack can be made on the seal. It may prove necessary to bring larger equipment to the platform if it is not possible to use the wind girder and the tank diameter is such that hand-held equipment will not reach all the way across the roof (see also pictures at the end of Chapter 5).

Example of a seal fire being attacked from the gaugers platform with a hand-held foam monitor.

Attack of a rimseal fire from stairway—see how much manpower is needed to deploy flexible hoses when there is no fixed dry risers.

- Attack the fire using large foam monitors. Monitors should not be considered as the primary method of attacking rimseal fires (although it is recognized that use of monitors from elevated hydraulic platforms has been successful in some cases). With monitors, there is always the risk of tilting the roof. Preplanning for rimseal fires should consider the provision of alternative response equipment.

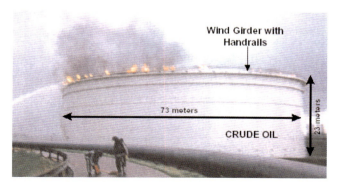

On tanks fitted with internal floating roof

Internal fires in these tanks are rare. However, when they occur, they are very difficult to tackle if the tank is not fitted with adequate fixed systems. Foam pourers combined with foam dams are the most effective and the design of these systems should be based on full surface fire.

Aluminium, pan type roofs, and open-top bulkhead pontoons should be assumed to sink and obstruct foam flow.

Ignition usually occurs during first filling with or without switch loading (refer to BP Process Safety Booklet *Safe Tank Farms and (Un)loading Operations*), exposure to radiant heat from a close-by fire or hot work.

Signs of an internal rimseal fire that ignited after this tank was exposed to radiant heat from another tank fire.

Fires at the vents are very difficult to tackle. This 95 ft (29 m) gasoline tank fire was extinguished by using foam and three of the vents were 'shot out' utilizing dry chemical (Hydro-Chem™ type).

7.3 Bund (dike) fires

The general rules for full surface fires are applicable here. The technique for fighting fires in bunded (diked) areas is to extinguish and secure one area then to move on to and extinguish the next section of the bund. This procedure is continued until the complete bunded (diked) area is extinguished.

Before extinguishing fires in a bund (dike) it is important to ensure that the flammable liquid remaining does not pose a greater hazard than if it had been allowed to continue burning (such as if the burning liquid is benzene). It will be necessary to keep this liquid covered with a blanket of foam.

Equipment requirements

NFPA 11 recommends fixed foam pourers for common bunds surrounding multiple tanks with poor access or less than 0.5 tank diameter spacings.

Minimum application rates for bund pourers are (from BS 5306):

- 4 l/min/m^2 (0.1 gpm/ft^2) for hydrocarbons;
- 6.5 l/min/m^2 (0.16 gpm/ft^2) for foam destructive liquids.

However, BS 5306 specifies that there should be one 2,600 l/min (690 gpm) discharge device (low or medium expansion) for each 450 m^2 (4,800 ft^2). Discharge time is calculated for 60 minutes.

This foam equipment should be capable of being operated simultaneously with tank-surface foaming operations. However, the bund fire must be extinguished prior to *attacking* the tank fires (if not, it will reignite the tanks).

There should be sufficient monitors and hand-held foam nozzles (or fixed systems depending on the manpower available) to deal with any bund fire that may occur. When applying foam on a bund fire with monitors, the tanks should be used as a deflector plate. This keeps the flame from the tank shell and starts the foam blanket where it is most urgently needed. The use of water monitors in tandem with foam monitors creates foam application dilution problems—only minimum cooling water should be applied when foaming.

There is value in keeping the water level above any product lines in the bund as this will protect them from the effects of radiated heat and flame impingement, and will help prevent the 'spreading' of flanges.

Use of medium expansion can be a very effective tool in quickly suppressing bund fires, as demonstrated by the following pictures:

Never commit firefighters into a bund contaminated with product, even if the spill is covered with foam. In a situation such as the ones illustrated in pictures below, wind or a water stream can open the foam blanket and fuel can reignite in seconds (see St. Ouen incident described in Appendix 9 of this booklet).

7.4 Foam supplies

The quantity required varies according to the tank size and the use of fixed or portable equipment.

A foam attack must be capable of being sustained for a minimum period. Therefore the quantity available before commencing the foam attack should reflect this requirement.

If the foam supplies are in drum storage then the logistics of supply will need to be considered. There will also be a need for additional manpower and equipment such as forklift trucks and vehicles to transport the drums to the area where they are to be used. Mechanical or manual transfer pumps may also be required.

Foam stored in bulk will require access. Mobile tanks may require towing vehicles. Care must be taken to ensure free access by foam vehicles for the duration of the incident.

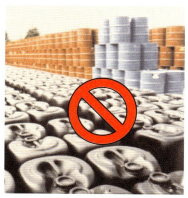

It is recommended that bulk storage is provided in mobile tankers, elevated bulk storage tanks or 1,000 litre (265 gallons) Schutz containers on mobile platforms. In each case pre-fire plans should ensure that immediate access to foam storage units is available 24 hours a day and appropriate valves, pipework or foam pumps are provided to decant the foam compound. The use of 25 or 200 litre (6.5 or 53 gallons) drums is not acceptable due to the intense use of manpower needed to mobilize and decant them.

6% foam concentrates shall not be used anymore for new installations. 1% foam concentrates are strongly recommended for all mobile equipment options for tanks of 70 m (230 ft) diameter or bigger as this concentration helps to reduce significantly the foam logistics. If seawater is used, application rates should be increased by 20% to take into account significant degradation of finished foam quality, unless detailed tests can give a more accurate value.

7.5 Water supplies

As water constitutes 97% at 3% concentration of finished foam solution, considerable quantities of water will be required for the production of foam for mounting an attack on tank fires, dike/bund fires and for the cooling of exposed equipment.

The following table *(given here as illustration for the specific case of a full surface tank fire supposed to be containing gasoline and using mobile equipment only, with 3% foam)* gives an indication of water required for foam production at the recommended rate of application in tanks of varying diameter—water and foam are mixed and applied at the NFPA +60% rate which equals 10.4 l/min/m² (0.26 gpm/ft²).

Tank diameter		Approximate rate of foam solution application rate		Water needed for foam production only (add cooling if needed)		3% Foam concentrate		Total foam concentrate required for 65 min application	
m	ft	lpm	gpm	m³/h	gph	lpm	gpm	litres	gallons
8	26	523	138	30	8 284	16	4	1 019	269
10	33	817	216	48	12 941	25	7	1 593	421
12	39	1 176	310	68	18 628	35	9	2 294	606
14	46	1 601	423	93	25 360	48	13	3 122	824
16	52	2 091	552	122	33 121	63	17	4 078	1 077
18	59	2 646	699	154	41 913	79	21	5 161	1 363
20	66	3 267	862	190	51 749	98	26	6 371	1 682
22	72	3 953	1 044	230	62 616	119	31	7 709	2 035
24	79	4 705	1 242	274	74 527	141	37	9 174	2 422
26	85	5 522	1 458	321	87 468	166	44	10 767	2 842
28	92	6 404	1 691	373	101 439	192	51	12 487	3 297
30	98	7 351	1 941	428	116 440	221	58	14 335	3 784
40	131	13 069	3 450	761	207 013	392	103	25 485	6 728
50	164	20 420	5 391	1 188	323 453	613	162	39 820	10 512
60	197	29 405	7 763	1 711	465 775	882	233	57 340	15 138
80	262	52 276	13 801	3 042	828 052	1 568	414	101 939	26 912
100	328	81 682	21 564	4 754	1 293 843	2 450	647	159 279	42 050
120	394	117 622	31 052	6 846	1 863 132	3 529	932	229 362	60 552

An example of foam/water requirements (NFPA +60%)

Note: To complement this book, BP Refining Fire Community of Practice produced two double slide rules to use for training purposes. These study plastic slide rules enable users to estimate the application flow, and foam and water quantities required for:

- *Slide 1 side 1: a full surface tank fire using mobile firefighting equipment.*
- *Slide 1 side 2: a full surface tank fire using fixed firefighting equipment.*
- *Slide 2 side 1: a rim seal fire using fixed firefighting equipment.*

- *Slide 2 side 2: a bund (dike) fire, based on the spill surface or bund surface on fire.*

The first three sides calculations are based on inputting the tank diameter.

These handy slide rules are valuable training tools for tank designers, industrial firefighters and fire brigade officers. Detailing 1%, 3% and 6% foam concentrates, they cover both Metric (SI) and Imperial units (US) and are available from IChemE Booksales (sales@icheme.org).

The water for foam production will need to be supplied to the one location so it will be necessary to evaluate hose requirements, along with intermediate pumps for water relays if the supplies at the chosen location are inadequate. Consideration should be given to the use of a large diameter hose for feeding foam monitors i.e. 152 mm (6 in) diameter or greater. A hydraulic study and/or practical exercise should be carried out at the pre-fire planning stage to ensure that adequate supplies at the appropriate pressure are available.

Pre-fire plans should consider **total** water consumption from both direct foam attack on the tank involved in the fire and adjacent risks from radiated heat needing water spray protection.

8
Conclusions

Factors that will increase the probability of successful extinguishment of a storage tank fire are:

- The use of low expansion-ratio aspirated foam.
- Adequate application rate.
- Large capacity water/foam monitors in sufficient numbers.
- Efficient handling of foam concentrates.
- Sufficient water supply to monitors (volume and pressure).
- Adequate time and manpower.
- No attempt should be made to apply foam unless sufficient resources are available to mount an extended attack for the recommended duration of application. However, the GESIP tests (see bibliography on page 45) show that, if there are sufficient foam stocks, an early continuous foam application at half the extinguishment rate is efficient in reducing the thermal flux, and therefore, in reducing the strain on firefighters and the probability of escalation. This, of course, relies on the concentrate being in good condition.

If it is planned to let a tank fire burn out then it is important for firefighters to know that tank shells exposed to fire normally fail by folding inwards above the liquid. Therefore the available water supplies should be directed to protecting exposures and not on the shell of the tank on fire. External roof drains on floating roof tanks which are normally left open should be closed to prevent the loss of flammable material into the bund.

Storage tank fires are often spectacular in nature, generating much heat accompanied by a highly visible column of smoke. However the application of the correct techniques has resulted in many such fires being successfully extinguished. If the foam is being applied correctly then visible evidence of fire reduction should be seen in less than thirty minutes after commencement. If no signs are seen then further checks need to be carried out to ensure that the correct rates are being applied.

Appendix 1: Short bibliography

- NFPA Codes, Standards and Journal
- GESIP foam tests reports
- LASTFIRE study, risk workbook and foam tests
- API 2021 last edition
- Fire Service Manual Volume 2 Petrochemical (The Stationary Office 2001 edition)
- Technica Report on the Singapore Tank Fire 1988
- BP fire school manual
- Resource Protection international 'foam seminar' documents
- BP booklet 'Alternatives to halon 1211 and 1301 fire fighting suppressants', May 2004 edition
- BP fire response workbook, 1994 edition
- BP booklet 'Fire Protective Clothing', May 2004 edition
- BP Engineering Technical Practices 44-10 and 24-40
- Model Code of Safe Practice in the Petroleum Industry: Part 19. Fire Precautions at Petroleum Refineries and Bulk Storage Installations, Institute of Petroleum, ISBN 04719 43282
- Face au Risque
- Industrial Fire Journal
- Industrial Fire World Magazine
- Fire International
- See also all references in BP Process Safety Booklet *Safe Tank Farms and (Un)loading Operations*
- INERIS report 13 on Boilover, March 2003
- 'Control of Major Accident Hazard (COMAH) Competent Authority Policy on Containment of Bulk Hazardous Liquids at COMAH establishments', UK HSE and EA, February 2008
- 'Storage of flammable liquids in containers, HSG 51', HSE Books, ISBN 07176 14719
- 'Storage of flammable liquids in tanks, HSG 176', HSE Books, ISBN 07176 14700

- 'Pollution prevention guidelines: Controlled burn; PPG 28', UK EA, July 2007
- 'Pollution prevention guidelines: Managing firewater and major spillages; PPG 18', UK EA
- 'Fire Systems Integrity Assurance' available from www.ogp.org.uk.

Most relevant videos

- SRC (Singapore) tank fire;
- Total St Ouen depot fire;
- Denver airport tank farm fire;
- Jacksonville fire;
- Neste Panva Finland fire;
- Sunoco Sarnia Canada tank fire.

Acronyms and abbreviations

API	American Petroleum Institute
AR	Alcohol Resistant
BS	British Standard
FSIA	Fire Systems Integrity Assurance
GESIP	Groupe d'Etudes et de Sécurité dans L'Industrie Pétrolière
IR	Infra Red
LHD	Linear Heat Detection
LPG	Liquefied Petroleum Gas (Propane–Butane)
NFPA	National Fire Protection Association
QSB	Quarterly Safety Bulletins
TVP	True Vapour Pressure

Appendix 2: Critical application

In any application of foam to a fire in a flammable liquid, there is a 'critical' application rate below which the fire cannot be extinguished. This is due to the heat of the fire and the flammable liquid destroying the foam before it can effectively cool the area to which it has been applied (see figure below).

There is an optimum application rate at which the fire can be extinguished with less foam than any other rate.

There is a preferred application rate which is safely above the 'critical rate' and is reflected in published application rates in codes and standards. This preferred rate will give a useful faster extinguishment time rather than the minimum rate while using only marginally more foam.

It is recommended that when tackling a fire in a flammable liquid, if no appreciable lessening of the fire intensity takes place within the first 20 to 30 minutes of the foam attack, the rate of application should be reviewed.

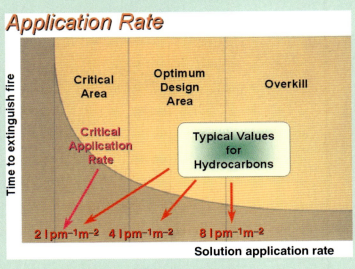

Critical application

Appendix 3: Escalation

In 1989–90 Technica carried out a study on behalf of the Oil and Petrochemical Industries Technical and Safety Committee in Singapore on the fire risks associated with atmospheric storage tanks. The following information is based on the Technica Report.

A full surface 50 m diameter open-top floating roof naphtha tank fire could be expected to escalate to fully involve a neighbouring identical tank in approximately 1.5 hours under the following conditions:

- 4 m/s (14 km/hr–13 ft/s) wind towards neighbouring tank;
- intertank separation of 0.5 diameter (25m–82 ft);
- neighbouring tank having pontoon roof and inadequate water spray protection.

Variants on the above base case give the following times for hazardous conditions to be generated at an adjacent tank, assuming in each case that all other parameters are unchanged.

Base Case	1.5 hr
Variants	
0.3 D separation	1.5 hr
1.0 D separation	3.0 hr
2.0 D separation	17.0 hr
No wind	2.8 hr
Water Sprays	2.8 hr
Double deck roof	1.5 hr
Water + no wind	4.5 hr
Water + double deck	>24.0 h

It can therefore be concluded that:

- escalation is likely for unprotected tanks of volatile material with normal separations unless the original fire is quickly extinguished;
- calm conditions only delay the escalation potential;
- moderately increased separation alone only delays the escalation potential;
- water spray protection or roof insulation alone does not prevent escalation;
- water spray protection or roof insulation are effective;
- smaller diameter tanks at normal separations are at greater risk of escalation than larger diameter equivalent tanks;
- lower volatility fuels allow more response time for firefighters.

NFPA 30 standard on tank location and bunding/diking should be regarded as a strict minimum and is insufficient to prevent escalation in case of fire. Sound design should provide large spacing in-between tanks (particularly those containing products susceptible to boilover: see Appendix 5) and between tanks and other process equipments. Good spacing also allows reducing the amount of water required for cooling.

Good spacing prevented escalation of this tank fire to adjacent equipment.

Example of poor design where elevated piperacks are exposed to radiant heat from both bund fires or tank fires, and would prevent easy monitor reach into tanks or bunds.

Appendix 4: Pre-fire plan checklist

This list does not seek to be exhaustive but will form the basis of all pre-fire planning. Any pre-fire plan must be verified by practical real time exercises and any deficiency taken account of in a modification to the plan. Whenever possible, exercises should be video recorded for use at subsequent de-briefs (see bibliography on page 44 for comprehensive advice on this topic).

Checklist

a) The type of tanks involved and their characteristics including:

- tank types, dimensions, contents, capacities;
- suction and discharge points, pipe diameters, lengths;
- pipe isolation valve locations, nearest, next nearest;
- isolation valve types, motor operated or manual;
- pipe capacities between valves, tanks and valves;
- product transfer capability;
- bund/dike dimensions (surface, height, slope profile . . .);
- bund drains, sewers and capacities;
- pipeways/piperacks locations;
- fixed fire protection systems, if any (foam, halon, etc);
- fixed water cooling system, if any;
- access roads;
- exposure risks (from radiated heat, direct flame, etc).

b) Availability of firefighting resources:

- portable/mobile firefighting equipment at risk area;
- portable/mobile firefighting equipment in reserve;
- trained manpower available (full-time, volunteer, etc).

c) Among the problems that should be looked for during this evaluation are:
- fire hose connections (compatibility, threads, etc);
- fire equipment (difference in types, operation);
- communications (difference in frequencies, channels);
- foam concentrates (difference in types, proportioning percentage etc);
- fire crews (full-time, volunteers, part-time, etc);
- shift systems (difference in shift schedules, reliefs);
- vehicle sizes (over-large for existing roads, bridges).

Additionally, the following support resources should also be identified and evaluated:
- specialist personnel (chemists, engineers, etc);
- communications equipment (fixed, mobile, frequencies, etc);
- transport/supply (supply trucks for reserve equipment);
- security (incident area, general area patrolling);
- medical (ambulances, medical staff, medical facilities);
- stores/warehouses (staff availability, normal/after hours);
- emergency alert system (alarm, siren, etc);
- fire water supply (volume, pressure, capacity);
- fire water pumps (types, quantities, locations, operation);
- hydrants (locations, discharge, outlet types);
- hydrant/monitors (locations, nozzle flow);
- foam concentrate storage for fixed foam systems;
- foam concentrate storage for portable/mobile equipment;
- foam concentrate volume on fire appliances;
- foam concentrate stocks in reserve;
- fire appliances (types, capacities, pumps, equipment).

Appendix 5: Specific hazards

A5.1 Boilovers

Boilover is the term given to the expulsion of burning oil from an open top tank involved in a full surface fire or a superheated tank involved in a bund fire. Large boilovers occurred at Yokkaichi (Japan) on 15 October 1955, Tacoa (Venezuela) on 19 December 1982, at Milford Haven (UK) on 30 August 1983, at Thessalonica (Greece) on 24 February 1986, at Port Edouard Herriot (France) on 2 June 1987 and at Skikda (Algeria) on 4 October 2005.

	Fire	Product	Time to boilover from start of fire
Yokkaichi	Tank fire	7,000m^3 of fuel oil	6.5 hours
Tacoa	Tank fire	14,000m^3 of fuel oil #6	6 hours
Milford Haven	Tank fire	47,000 t of crude oil	12.5 & 15 hours
Thessalonica	Tank fire	17,900m^3 of crude oil	30 hours
	Tank and bund fire	10,350m^3 of fuel oil	5 days
Edouard Herriot	Bund fire	1,000m^3 of diesel oil	2.5 hours
Skikda	Tank fire	35,000m^3 of crude oil	14 hours

A boilover occurs when hot residues from the surface of the burning liquid become denser than the unburnt oil and sink below the surface to form a hot layer which sinks much faster than the level of liquid drops due to the rate of burning.

As it sinks towards the bottom of the tank, the heat layer increases in size and density with temperatures in the range of 300 to 600°F (150 to 320°C).

When this heated layer reaches a water or water emulsion layer, it first superheats the water then causes a steam explosion. The water flashes to steam at temperatures in excess of 212°F (100°C) and will expand by as much as 2000 to 1.

It is estimated that this steam explosion can propel burning oil and vapour to a height of ten times the tank diameter.

To be liable to boilover, oils must have components containing a wide range of boiling points. Most crude oils fall into this category.

The three elements necessary for a boilover are:

- a fire in an open top tank, involving all or most of the surface; or a tank fully involved in a bund fire;
- a layer of water or water-oil emulsion in the tank;
- the development of a heat layer which is determined by the properties of the stored material.

When a fire in an open top tank containing a flammable liquid with a wide range of boiling points, such as crude oil, begins to burn, the components with the higher boiling points sink below the surface and form a heavy heated layer. In some crude oils, this layer travels downward into the oil at only 7 cm (3 in) per hour, while in others it may be as much as 2 m (78 in) per hour. In most crudes the rate is from 30 to 45 cm (12 to 18 in) per hour faster than burnoff.

Pre-fire plans should take into account the fact that pumping out a tank on fire that contains a boilover fuel may bring forward the time when a boilover would occur. However, the amount of fuel involved in the boilover would be reduced by pumping out.

Additional signs of a potential boilover are:

- an increase in flame height and brightness;
- a change in sound to crackling or frying;
- blobs of burning material may be ejected a few metres from the tank.

Historical note: In the early days of the Oil Industry, some storage sites in the US used to have a Civil War era gun, to fire a solid shot through a crude oil tank wall that was on fire. This would spill the content of the tank in the bund and avoid a boil-over.

ACCIDENT On 19 December 1982, in a power plant in Tacao near Caracas (Venezuela), a huge boilover (see explanation below) occurred on a fuel oil tank, killing at least 160 people in a huge fire ball.

The installation comprised three power plants (one under construction) by the sea side, and a tank farm on the hill above. The site was surrounded by a poor residential area.

On 18 December, operators transferred fuel oil #6 from tank 9 to tank 8, on top of the hill (see drawing below giving approximate distances and elevations). Tank 8 was a 40,000 m^3 fixed roof tank of 55 m (180 ft) diameter, filled approximately with 14,000 m^3 of fuel oil.

(continued)

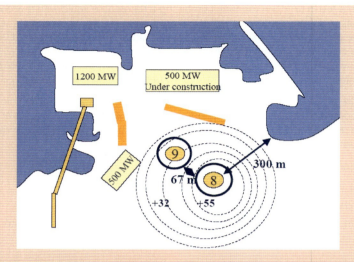

At 23:30, seeing that the product temperature was way too high (88°C/190°F instead of 65°C), operators cut a tracing system to allow temperature to cool off. The next morning before dawn (around 06:00), two operators went to manually gauge the tank. During the operation, an explosion occurred, either due to the use of non-intrinsically safe lamps or to a static spark. The temperature of the product was still above 80°C/176°F, well above the flash point of 71°C/160°F. The roof of the tank landed some distance away and severed some product lines, igniting a bund fire. The roof also supported the foam systems which were ripped off when the roof took off. Tank 8 immediately suffered a full surface fire.

Fire fighting action was limited by the remoteness of the site, inadequate access to the tank (hilly terrain and the only access road was below the bund on fire) and damage caused to the fixed fire protection systems. Fire fighters, civil defence personnel, plant workers, journalists and onlookers were within 30 to 60 m (100 to 200 ft) of the tank.

At 12:20, a massive boilover occurred, with a fire ball approximately 150 m (500 ft) diameter, raising up to 600 m (2,000 ft) high, and burning product fell down around the site.

The boilover pushed product over the top of the tank, creating a wave of burning liquid that went over the 6 m (20 ft) high bund wall and submerged vehicles and people alike and travelled more than 400 m (1,300 ft) downhill. This wave entered another bund and ignited another fuel oil tank that will burn for many days. Tank 8 fire was extinguished by the boilover.

(continued)

Tank 8 is in the foreground, tank 9 is in the middle of the picture.
Note the steep hill where the tanks are located and unique access road below.

This incident killed at least 160 people including 40 firefighters and eight journalists, injured more than 500, destroyed 60 vehicles and 70 houses and most of the power generation plant (1200 MW and one 500 MW section destroyed). 40,000 people where evacuated by the army as long as another boilover was feared from the other tank on fire.

A5.2 Rocketing tanks

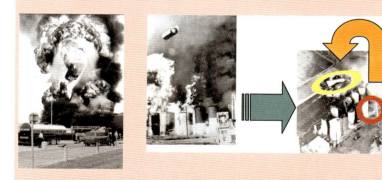

Rocketing tanks are always small diameter tanks (mostly less than 12 metres diameter) because the weld of the roof is not frangible. Overheated small tanks with low levels of products are most susceptible to rocketing.

A5.3 Tank failure

Tank failure is a rare occurrence but can happen with or without fire. The failure can be partial (such as the roof blown away) or catastrophic. Emergency plans should take this risk into consideration, in particular, as far as reasonably possible, by not locating firefighters 'downhill' from exposed tanks. For this risk and also to limit exposure to radiant heat, it is good practice to favour long range firefighting equipment that does not require continual readjustment or continuous manning for operation.

Appendix 6: Properties of foams and other extinguishants

A6.1 General

To have a good-quality foam blanket, the water quality must be good. Water may be hard or soft, fresh or salt, but it must be of suitable quality so that it does not have an adverse effect on foam formation or foam stability. Corrosion inhibitors, emulsion breaking chemicals, or any other additives must not be used without consultation with the foam concentrate supplier. Recycled water from skim ponds or separators is generally not acceptable because trace amounts of oil can affect the quality of the foam blanket.

The fire will not be extinguished unless the foam is continuously applied at the recommended application rate for the specified minimum time. A sealing, cohesive blanket of foam must be established and maintained over the surface of the liquid. The integrity of the blanket should be monitored and maintained for the safety of personnel and to ensure extinguishment.

Of particular note is the use of freeze protective additives. The addition of substantial percentages of glycols can lead to the foam concentrate exhibiting flammable properties as the glycols vaporize.

Environmental considerations with respect to foam (metal salts, stabilizers, fluoro-surfactants, solvents, preservatives etc., all have environmental effects) should be reviewed before buying any new foam. This is particularly important as environmental concerns on the use of fluorosurfactants, one of the main ingredients of modern firefighting foams, are tending to increase. Also, issues like oxygen demand or lethal concentrations should be addressed, particularly if the foam may be used near water courses.

In areas with specific environmental concerns, it might be important to gain local Environmental Authority and Fire Brigade formal approval on the type of foam to be used before an emergency occurs.

Note: In 2002, some manufacturers were developing fluorine-free foams but none had been fully tested under LASTFIRE tests.

A6.2 Proteins and chemical foams

Proteins were introduced in 1935 and whilst they have excellent heat resistance and stability they lack fuel tolerance and have slow knockdown performance.

Together with chemical foams, which contain alkaline and acidic salts, standard protein foams are not generally recommended.

A6.3 Fluoroprotein foam (FP)

FP concentrates are formed by the addition of synthetic surfactants to a protein foam. These foams:

- are generally compatible with dry powder (so can be used in conjunction with dry powder to fight 3D fires with the proper equipment and training);
- give rapid knockdown compared with protein;
- have good fuel tolerance and resistance to fuel entrainment;
- can withstand rougher application by hose streams;
- are suitable for sub-surface injection for non-polar liquid hydrocarbons;
- have good stability and burnback qualities.

In general fluoroprotein foams are a cost effective or general purpose foam suitable for the majority of installations. The recommended concentration for these foams is 3%.

Good FP foams generally show less performance degradation than others when used with sea water.

A6.4 Aqueous film forming foams (AFFF)

AFFFs are a combination of fluorocarbon surfactants and synthetic foaming agents that act as a barrier to exclude air and form a vapour-sealing aqueous film on a hydrocarbon surface. AFFFs generally:

- are compatible with other foams if generated separately;
- are available at 1 to 6 percent concentration;
- are compatible with dry powder;
- are more fluid than other foams;
- give rapid knockdown and initial fire control;
- tend to give less stable foam than FP and consequently less burnback resistance.

Because of their film forming ability and their low energy requirement to produce good quality foam, AFFFs can be used through non-aspirating equipment such as conventional sprinkler heads. Thus, existing water deluge systems can be easily converted to foam systems by merely adding the appropriate proportioning equipment (for example, to fight a bund area fire and keep fire away from the tank shell).

AFFFs also act as excellent wetting agents and hence are gaining in popularity in multi-purpose hand-held extinguishers.

Standard AFFFs are not generally recommended for storage tank application.

A6.5 Synthetic detergent foam (Syndet)

These foam concentrates are based on a mixture of synthetic foaming agents with additional stabilizers. They are versatile in that they can be used to produce low, medium or high expansion foam. However, they have found little acceptance in the petrochemical world because they do not exhibit very good burnback resistance or fuel tolerance.

A6.6 Film forming fluoroproteins (FFFP)

FFFPs possess a combination of AFFF and FP characteristics having the suppressant vapour forming aspect of AFFF and the sealing quality against re-ignition of the FP foams.

The good resistance to fuel entrainment makes FFFP suitable for both over-the-top and sub-surface injection for hydrocarbons fires.

The FFFPs developed for use with polar solvents normally require a higher concentration on such liquids than when being used on ordinary hydrocarbons.

Because FFFPs are generally two or three times as expensive as other foams, some installations will only consider them where the specific nature of the risk warrants their use.

A6.7 Alcohol resistant foam concentrates ('multipurpose' types)

Polar solvents and water miscible fuels such as alcohols are destructive to standard hydrocarbon type foams because they extract the water contained in them and rapidly destroy the foam blanket. Therefore these fuels require a special 'alcohol resistant' concentrate.

These foams can be synthetic or protein-based and are produced from a combination of stabilizers, foaming agents, fluorocarbons and certain special additives. The additives remain in the foam until it comes into contact with the polar solvent. As the polar solvent extracts the water in the foam blanket they form a polymeric membrane which prevents the destruction of the foam blanket.

On a hydrocarbon fuel this foam acts as a conventional foam. Hence it forms an effective agent for both types of flammable liquid. The AFFF based multi-purpose concentrates produce foam that tends to be more stable than their standard AFFF counterpart. Hence they are the synthetic concentrate of choice for tank incidents. When designing systems for water miscible fuels it is important to consult the foam manufacturer regarding the correct application rate for a particular fuel prior to finalizing any foam requirements.

The recommended concentration for these foams is 3% (or 1% with adequate proportioning and mixing equipment) for hydrocarbons fires; and 3 or 6% for water soluble fuels.

ACCIDENT This 5,000 m³ fixed roof tank contained 1,000 m³ of pure ethanol when it was struck by lightning. The fire was extinguished in 3 hours. As refineries use more and more alcohol based products to improve octane level or to produce bio-fuels, it is important to note that these fires require greater quantities of foam concentrate than hydrocarbon fuel fires: 23,000 litres of foam concentrates and 7,000 m³ of water were used during this incident. When designing a new alcohol storage, it is crucial to check that the foam and water systems are adequate and that the surfaces that can potentially be on fire (tank, bund, etc.) are minimized.

Also note that this type of fire produces very little smoke and sometimes, even the flames are invisible, except with Infra Red (IR) imagery.

A6.8 Dry powder

Whilst dry chemicals can be very efficient at securing quick knockdown of hydrocarbon fuels they will not, by themselves, secure the fuel against reignition. At least seven different types of dry chemical are available on the market and care should be exercised to ensure compatibility with whatever foam stocks are being utilized.

The use of dry chemical together with AFFF can be particularly effective against three dimensional pressure or leak fires and spill fires with the proper equipment and training. The dry chemical extinguishes the fire and the foam seals the associated spill fire.

A6.9 Halogenated hydrocarbons

These are no longer recommended for use in hydrocarbon tank firefighting due to the adverse environmental effects of Halon 1211, 1301 and 2402. Halon alternatives are likewise discouraged: Fire detection is the preferred option—see A7.6.

When decommissioning halon systems, consideration must be given to the fire detection system to have an as early as possible alarm, and the foam system in place to fight rimseal fire (fixed system + foam connections on the gaugers platform level to allow the use of hand held foam nozzles).

Refer to bibliography for more details.

A6.10 Testing procedures for foam systems and foam concentrate

To demonstrate that a foam system can be expected to function effectively when called upon, an operator ought to be able to show that a system of foam system and foam concentrate testing is in place within a framework of FSIA (Fire Systems Integrity Assurance: available from www.ogp.org.uk).

The most obvious requirements in the development of a comprehensive in-house routine testing procedure are:

- defined test intervals;
- precisely defined and documented testing methods;
- specific acceptable values of test parameters;
- documentation to record results;
- review procedure.

Test intervals

The standards may recommend a suitable test and inspection schedule as follows:

- **Weekly:** Test and check pumps (including fuel levels and control position (automatic/manual)) and that the water supply is available at the right pressure. Check foam levels.
- **Monthly:** Visual check that there are no leaks or obvious damage to pipework, and all operating controls and components are properly set and undamaged.

  Operation of all equipment located in a 'marine' environment (such as monitors, valves).
- **3 months:** Testing and servicing of all related electrical detection and alarm systems. (More detailed test requirements are provided in standards relating to fire detection).
- **6 months:**

  Foam producing equipment

  Inspection of proportioning devices, their anciliary equipment and foam makers for mechanical damage, corrosion, blockage of air inlets and correct manual function of all valves.

Pipework

Examination of above ground pipework to determine its condition and that proper drainage pitch is maintained. Hydraulic pressure testing of normally dry pipework when visual inspection indicates questionable strength due to corrosion or mechanical damage.

Strainers

Inspection and cleaning of strainers. (This is essential after use of the system and after any flow test.)

Valves

Check of all control valves for correct manual function and automatic valves additionally for correct automatic operation.

Tanks

Visual inspection of all foam concentrate and foam solution tanks, without draining. Checks of shipping containers of concentrate for evidence of deterioration.

- **12 months:**   Laboratory test of foam concentrate or solution for changes in constitution or characteristics and the formation of sediment or precipitate if more than three years old.

  Check that all personnel who may have to operate the equipment or system are properly trained and authorized to do so, and in particular that new employees have instruction in its use.

  Discharge test of foam equipment (top pourers).

- **5 years:**   Fire test of foam concentrate (to be done also on purchase of each large batch).

- **As required by statutory regulations:**   Inspect internally all tanks.

Note: Records must be kept of each check/test results, along with pictures of the main tests. *It must be emphasized that these are only general recommendations and should be developed to suit a particular system but they do provide some helpful guidance for most systems.*

Top pourers foam test on a cone roof tank (good pourers are provided with a curved pipe that sends foam away from the shell during tests).

Test of a medium expansion foam generator for bund protection

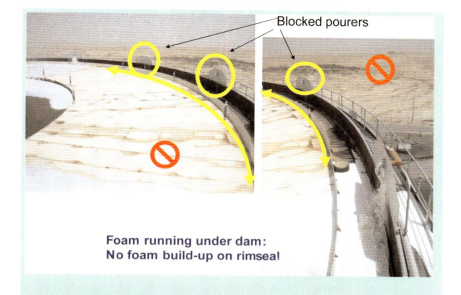

Foam running under dam:
No foam build-up on rimseal

Defining and documenting test methods

Any foam equipment, either portable or fixed, should include descriptions of detailed testing methods. In practice the documentation provided to operators is often very poor and consists of a few data sheets on system components put together as a 'manual'. At the very minimum, the documentation should include step-by-step instructions of how to measure the system parameters described in standards such as NFPA 11 (i.e. systems flow, time to achieve effective discharge, proportioning rate, expansion and drainage time.)

A6.11 Foam system tests

Foam sampling

The ultimate test for the system hardware is to carry out tests on the proportioning accuracy and finished foam properties. The tests that can be carried out in the field during commissioning and at subsequent routine intervals are:

- foam expansion;
- drainage time;
- application rate;
- solution strength (Proportioning Accuracy).

Foam expansion and drainage time are sometimes referred to jointly as 'foam quality'.

Fire testing

A regular fire test is essential to find out the true capability of any foam concentrate—afterall, its ultimate purpose is to prevent or extinguish a fire.

Several fire tests (see picture below) have been developed around the world. Some are good and selective, others very poor allowing low quality foam to pass. All have been designed with a particular risk or foam concentrate in mind. It is quite possible none of them test the precise properties for a particular application. In addition, most of them are only of the pass/fail type, so they do not usually differentiate between several foams that meet a minimum requirement.

When evaluating fire performance of foam for storage tank application, a 'LASTFIRE' Foam Test For Storage Tank Fires should be specified for each large bulk purchase and then every five years when large stocks exist and each time a doubt exists on foam quality (regarding contamination or storage issues, for example).

Appendix 7: Firefighting equipment

A7.1 Fire pumps

Firewater systems (pumps, piping and hydrants) should be based on the credible 'worst case' fire water demands.

The fire main should be pressurized at all times by jockey pumps maintaining 4–6 bars pressure at all times with an automatic start of main pumps, in sequence, on pressure drop/flow take-off.

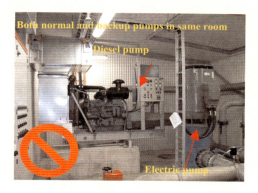

Where the fire water supply is obtained from static storage such as a tank or reservoir, then the reserve for firefighting purposes should be equivalent to the needs of the scenario-based assessment. (Some standards demand a fixed running time, for example six hours, at the minimum flow rate. It is better to review the 'design scenarios' and provide water according to them. For example, a controlled burn down policy may require in excess of this.)

Separation of fire pumps from hazardous areas should ideally be determined using a risk based approach. Some plant layout codes may prescribe a distance, for example, 100 m (330 ft). Fire pumps should in any case be protected from the effects of blast and/or thermal radiation from a fire. It is common to have multiple pump arrangements divided between two well separated pumphouses to protect against a common mode situation.

A7.2 Fire main and hydrants

Fire mains within a facility should be designed as a grid with isolating valves to allow maintenance without reducing cover to major exposures. Automatic vents should be in place at high points. Fire main material should be adequate to protect against the corrosivity of the water used (GRP is recommended for underground piping), but care is needed with joints.

The minimum configuration for a hydrant is 2×65 mm (2.5 in) outlets $+ 1 \times 150$ mm (6 in) outlet but can be much bigger if high capacity mobile equipment is required. Hydrants should be spaced at intervals of not more than 45 m (150 ft) in the hazardous areas and not more than 90m along the approach or access routes. Hydrants should be readily accessible from roadways or approach routes and located or protected in such a way that they will not be prone to physical damage.

A7.3 Fire water source

As far as practicable, unlimited water sources (sea, river, canal, etc) should be preferred to a storage tank or small lagoon solution. Again, it is good practice to have multiple source arrangements, well separated, to protect against a common mode situation.

This fire water tank was hit by multiple debris when an explosion occurred on a nearby process plant.

A7.4 Foam concentrate

1% or 3% are the most appropriate concentrations for storage tanks incidents. In areas with specific environmental concerns, it might be important to gain local Environmental Authority and Fire Brigade formal approval on the type of foam to be used before an emergency occurs.

Preference should be given to the supply of foam concentrate in bulk rather than in 25 or 200 litres (6.6 or 53 gallons) drums. The transport of drums is labour-intensive and time consuming. The use of either mobile tankers or large containerized packs of foam on a trailer with in-built supply piping and valves is a better arrangement.

Foam containers are better than drums but should be on a trailer for immediate use

The storage must be protected from heat, direct sunlight and cold. Recent LASTFIRE tests have proven that storage is a key factor in foam performance.

Foam quality must be tested yearly. This can be easily done on-site with the correct equipment and training, with reference to a specialized laboratory if on-site testing gives cause for concern.

A7.5 Fixed (cone) roof tank protection sytems

It is recommended that fixed roof tanks are fitted with a fixed top pourer foam system.

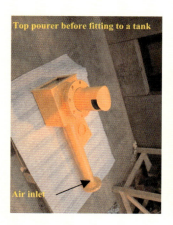
Top pourer before fitting to a tank
Air inlet

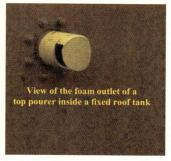
View of the foam outlet of a top pourer inside a fixed roof tank

During design, attention must be paid to the future maintenance and testing of top pourers (including factors such as access, weight of cover, foam projection on tank shell).

Note: Usually cover should be on top with easy access, but here there is good access from the staircase.

Bad example of top pourer with no access for test or maintenance

Note: Some 1960s foam pourers were designed to receive finished foam (air + foam concentrate + water) directly from ground level (see pictures below). These systems produce poor quality foam with very low expansion. They should be replaced with traditional pourers that receive only foam premix (foam concentrate + water) and inject the air at the top of the tank.

Air injectors to be connected at ground level after foam concentrate and water have been mixed.

Test of foam pourer—note poor quality of foam.

A7.6 Floating roof tank protection sytems

The tank rimseals should be fire retardant. The tank roof must also be fitted with a foam dam as specified in NFPA codes.

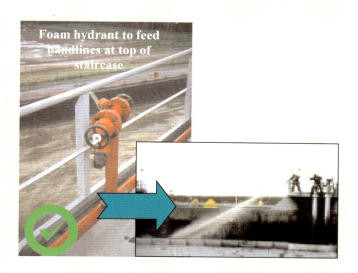

It is recommended to install handrails on the wind girder and two 65 mm (3 in) hydrant connections on the foam line near the gauger's platform with sufficient lengths of hose of 45 mm diameter and hand nozzles stored in a box at the top of the stairway, to facilitate a foam attack on a rimseal fire from the wind girder (for example, if a foam pourer is deficient). As stated previously on large diameter tanks, it is good practice to have additional outlets around the walkway circumference.

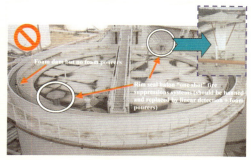

BCF (halon) systems must be banned and replaced only by fire detection (see below).

In many cases it is possible and justified to design foam top pourers to cover the whole of the tank roof area as well as the rimseal area, thus allowing for rimseal fires and full surface fires. This is usually relatively straightforward at low or no additional cost up to 35 m (115 ft) diameter tanks. Above that it might be a wiser approach to adopt foam pourers for a rimseal fire attack, and large mobile equipment for a full surface fire (see below).

Central foam distribution from a central foam manifold supply located on the roof (same principle as a roof drain but flowing from the bottom to the top) is not recommended as it relies on the roof being stable and afloat and any problem below the roof requires taking the tank out of service for repairs.

A7.7 Fire detection systems

Rimseal areas should be equipped with linear heat detection on floating roof tanks to give early warning of any small fire in that area. Electrical cable types are recommended for linear heat detection as they have a much more rapid response than pneumatic systems and are easier to test and maintain. Detail specification of all components including junction boxes, interconnecting cables, etc. to ensure minimum maintenance requirements.

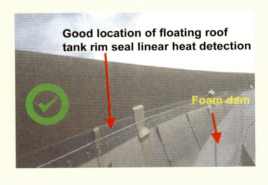

Good location of floating roof tank rim seal linear heat detection

Poor location of floating roof tank rim seal linear heat detection

Infrared cameras can be very useful during and after storage tank fires to locate hot spots (see picture below).

For process area, loading gantries, pump rooms, etc. the alternative of micro-CCTV detection or CCTV cameras linked to a fire detection software package are more and more being installed to give visible and real time pictures of an alarm to local or remote control rooms. With minimum manning on sites, such systems greatly increase emergency response.

A7.8 Mobile equipment

Sites where tanks are bigger than 30m (100 ft) diameter should pay very detailed attention to the equipment, training and manpower required if they choose the mobile equipment strategy as the preferred option. Today, the leading strategy is to use large equipment—the smaller the monitors the closer you have to get to the fire, so putting fire crews in an unacceptable risk situation. The smaller the monitor the more will be needed and the bigger the hose 'spaghetti' problem becomes (see picture below).

A7.9 Resource requirements for larger (>40 m/130 ft) diameter tanks

For ground level foam attack on a full surface tank fire, typical portable equipment includes the following:

- water monitors;
- water pumping appliances;
- large capacity foam monitors;
- foam pumping appliances;
- foam concentrate tankers or containers;
- fire hose including large diameter/capacity hose;
- water supplies.

All of the above points will be discussed as a practical review of the issues involved in a full surface tank fire and are included as a useful guide for equipment selection. It must be remembered that fire attack on large diameter tanks in the way described here has not always been successful, although there have been a few significant successes where the pre-planning and exercising and resource provisions greatly contributed towards the success.

Water monitors

Water monitors may be necessary for cooling adjacent tanks affected by radiant heat, or in some cases where tank spacing is inadequate, from flame impingement.

It is generally accepted that water cooling of the tank on fire is not normally necessary except, possibly, to assist foam blanket sealing against a hot tank wall. In some cases, cooling of the ignited tanks by monitors is thought to have led to distortion and consequent tank failure due to the creation of some cool sections of metal and some hot sections. All current evidence points to tanks folding inwards under full surface fire conditions. However, the behaviour of older tanks such as riveted tanks is uncertain. Therefore, the individual in charge of the fire attack must be responsible for deciding if and when cooling of the tank on fire should be carried out based on an assessment of the potential damage to the tank, the need for cooling to help the foam blanket/tank wall sealing, availability of cooling evenly around the tank circumference and any potential water drainage problems.

An important point to remember when using water or foam monitors is that water misting or drift will occur if there is any appreciable wind. If there are nearby power lines, this water may conduct electricity. Care needs to be exercised if there are any power lines in the vicinity of the tanks, or plant, to be cooled.

Water pumping appliances

The capacity of any water pumping appliances to be used for water monitor supply must be as large as possible. Typical 'standard' water tender/pumpers may have only a 2500 lpm (660 gpm) pump on-board. This will obviously only be enough for typical monitors, and if larger water monitors are to be used then two or more water pumping appliances may be required for every monitor.

This can lead to major logistical and deployment difficulties.

Examples of large mobile pumps

Fixed water monitors on fire vehicles

It is generally accepted that fixed water monitors on fire vehicle roofs or on hydraulic platforms or aerial ladders will only be of limited use during a full surface fire. The restricting factors in their use will be road access around the tank and distance to the tank from the safe parking area. It is also accepted that there is limited flexible use of the vehicle once it is parked and connected to hydrants. In other words, it may become a very expensive fixed monitor instead of a flexible response which can be moved around to suit circumstances.

Foam monitors

It is important to note that most recognized standards (such as NFPA 11) state that monitors should not be used as the primary attack method for tanks greater than approximately 18m (60 ft) diameter. However, in practice they have been used for much larger tanks, up to more than 80m diameter, although experience on tanks greater than 40m (130 ft) is limited.

If fire extinguishment is attempted, the importance of foam monitor capacities and stream range becomes obvious. Foam streams have to be such that the bulk of the output reaches the tank liquid surface.

It is recognized that the best method of application is to project foam with the wind behind the stream and not against. However, there have been, and will be situations where cross winds or a variable breeze causes reductions in stream ranges and these need to be considered.

It may be that a desktop calculation shows the range and trajectory of a foam monitor placed on a bund wall reaches the liquid surface easily only to discover in practice that the stream falls short due to a breeze or greater wind speed.

It is possible to supply foam to foam monitors by either using foam pumpers or foam pumps to create the water/foam solution or to use water tenders or large capacity water pumps to pump water to the foam monitors where foam concentrate is picked up via the monitor package induction. Both present logistical problems with the monitor induction method creating the greatest problems in terms of access to and around them.

A very important point to note is that when using foam monitors for full surface fires there will be losses from the foam stream due to thermal updrafts from the fire preventing some of the foam reaching the liquid surface. There will also be some loss due to stream feathering or fall out. With this in mind, a much higher total application rate is necessary to ensure that 6.5 l/m/m^2 (0.16 g/m/ft^2) reaches the fuel surface. It is now generally accepted that a foam solution production rate in the order of 10.4 l/m/m^2 (0.26 g/m/ft^2) or more should be used for foam monitor application on a full surface fire.

This high application rate often makes such application methods impracticable from an existing facility ring main.

Very large throughput monitors up to 60,000 l/min are now available as shown in the pictures below. It is obvious that such high flow rates cannot be met by typical local authority or industrial fire vehicles. Therefore, mobile firewater and foam pumps need to be part of the response package.

Example of a non-aspirated monitor with large mobile pump

Example of a large aspirated monitor

Foam pumping appliances

Foam pumpers for use at full surface tank fires would typically be used for supplying foam monitor flowrates of a minimum 5,000 l/min (1,320 g/min). Foam tank capacity onboard the foam pumper should be a minimum 5,000 litres (1,320 gallons). This would give over 30 minutes supply time to a 5000 l/min

(1,320 g/min) foam monitor, allowing time to replenish the onboard foam tank by either tankers or another method. Note that this is based on a 3% ratio. Obviously, if 6% is used, the time would be reduced to 15 minutes only.

If foam pumpers are to supply foam monitors with both foam and water then the onboard water pumps and foam proportioning systems should be capable of a minimum 5000 l/min (1,320 g/min) supply.

Note: Before purchasing any large monitor, real tests should be carried out to check adequacy of both monitor and water supply. On the day of a fire, it might be impossible or too hazardous to find a crane to lift the monitor into the bund as in the picture below:

Foam compatibility

Foam compatibility is an important factor. If mutual aid is to be used, a single foam concentrate at a uniform induction rate is preferred. Although it is possible, given suitable foams, to use, for example, AFFF-produced foam on a fire and then use fluoroprotein-produced foam on top of the AFFF blanket without any serious adverse effect, this is not recommended for large tank fires.

Mixing foam concentrates of different types is also not recommended and can completely destroy foam making capability. The intention should be to have a standardized foam type suitable for the fuel at a uniform induction rate so that there are no pump operator errors in proportioning.

Foam concentrate containers and supply considerations

Bulk movement and foam monitor supply of foam concentrate represents a major logistical problem which, if not carefully considered, will greatly delay foaming operations and, in some instances, will prevent effective and continuous foam application. It must be remembered that once foam application commences onto a tank surface fire, it must be maintained, uninterrupted, at the required rate for the duration required.

The typical methods of foam concentrate re-supply are:

- **25 litre (6 gallons) drums**

  The use of 25 litre (6 gallons) foam concentrate drums is theoretically possible during a large storage tank fire but very difficult. Using the example of the 5,000 l/min (1,320 g/min) foam monitor and a 3% induction rate, this monitor would use 135 l/min (36 g/min) foam concentrate, or more than five drums each minute which will probably result in interrupted supply and is very labour intensive. This is not considered a practicable option.

- **200 litre (53 gallons) drums**

  One 200 litre (53 gallon) drum of 3% foam concentrate supplying a 5,000 l/min (1,320 g/min) foam monitor would last for just under 1.5 minutes. With monitor flowrates above 5,000 l/min (1,320 g/min), the 200 litre drums will be used up in similar fashion to that of the 25 litre drums. For example, a 7,500 l/min (1,980 g/min) monitor would need 225 l/min (60 gallons) foam concentrate, less than 1 minute's supply using a 200 litre (53 gallons) drum. Again, this would create massive logistical problems.

- **1,000 + litre (265 gallons) containers**

  Large capacity 'polytank' containers of 1,000 litres (265 gallons) or more are an option for supplying large capacity monitors. They can be transported to each fire vehicle or monitor and dropped off on the spot within reach of the foam suction hose of the vehicles or monitors.

  Dropping several within suction hose reach will obviously increase the duration before changeover is necessary and therefore give more time to transport crews to keep foam concentrate re-supply moving but may cause congestion in an already restricted area. Trailer options are cheap and easy to maintain.

- **Foam tankers**

  Using foam tankers in the range of 10,000–15,000 litre (2,500–4,000 gallons) capacity is the preferred method of supply and re-supply for large capacity foam monitors, especially those of >15,000 l/min (4,000 g/min).

  Tankers can be also be used for refilling foam pumper onboard foam tanks, foam monitor trailer tanks or other foam containers at the monitor locations.

Fire hose/water delivery hose

The typical size of delivery hose for water monitors and foam monitors will be 70 mm (3 in) diameter. Usually these will be in 20–25m (65–82 ft) lengths.

Although there will be a need for the 70 mm (3 in) size, especially for water monitors, the use of large diameter or large flow hoses offers a less labour intensive option for deployment. Typically, the large diameter hose will be used from either hydrants or direct from large capacity fixed or mobile pumps. Sizes will vary depending on the capacity of the foam monitor or monitors.

The sizes of large capacity hoses will typically be:

- 100 mm (4 in);
- 120 mm (5 in);
- 150 mm (6 in);
- 250 mm (10 in);
- 300 mm (12 in).

Whilst recognizing the advantages of large diameter hoses it should be borne in mind that they may require special mechanical handling facilities due to their weight. Also, if only large hose is used, there may be no flexibility for combating other types of fires in a facility where smaller monitors are to be used which only require 70 mm (3 in) hose.

If hydrants are located very remote from the incident, it may be necessary to use hose trailers or hose layers rather than having a totally manual deployment.

Example of large delivery hoses

Water supplies

The water supply for firefighting large storage tank fires is the key to any fire response decision. Opting to combat a full surface fire requires full consideration of the existing water supplies beforehand. On at least two occasions it was discovered that the cooling of exposures plus the foam attack water requirements greatly exceeded normally available water supplies in terms of flow and pressure.

For example, if the tank is in the order of 80 m (260 ft) diameter, the total foam solution application rate for aspirated foam on a 3% induction based on 10.4 l/m/m^2 (0.26 g/m/ft^2) will be approximately 52,276 l/min (13,800 g/min) of which more than 50,700 l/min (13,400 g/min) will be water.

Add to this possible exposed tanks cooling based on 6 × 2500 l/min water monitors and the water rate required will exceed 67,000 l/min (17,700 g/min).

Pressures and flows of firewater systems are the cornerstone of any successful firefighting operation. In some cases reliance is placed on using pumpers or trailer pumps to draft from an open water source. Although this method may be successful, the logistics, in terms of vehicles, hose and manpower will need very careful coordination and supervision.

This particular water supply method also presents heavy maintenance demands for fire vehicles or trailer pumps.

An additional factor to consider is that any contaminated cooling water, or water/foam mixtures may need to be contained and treated prior to being 'released' into water streams or rivers etc.

Full surface fire portable equipment selection considerations

- Water monitor stream ranges are very important. The stream range (straight stream or jet) length and height (trajectory) of a water monitor as advertised by manufacturers will always give best possible figures obtained and it will always be under still air conditions. This is the only way to standardize the range figures. Therefore, end-users must consider their own particular typical weather conditions and winds to select appropriate monitors. The best method is to request or conduct tests of monitor stream ranges from anticipated positions to the tanks or plant in question under different wind conditions.

- The capacities of portable water monitors for cooling large diameter tanks should be considered from 2,500 l/min (660 g/min) upwards. It is usually the case that the smaller flowrates will not provide the desired range but it should also be noted that if higher capacity water monitors are to be used they will require higher capacity water pumping appliances.

- Water pumping appliances may be needed for supplying water monitors to cool exposures during tank firefighting. If the pumpers have limited capacity, say of only 2,250 l/min and the minimum size of water monitor to be used is 3,600 or more then obviously more than one pumper will be required for supplying one monitor.

Pumpers should have a minimum 4,500 l/min pumping capacity. If the site firewater ring main has adequate flow and pressure, then water tenders may not be required for monitor supply.

- In selecting foam monitors there are several important factors to bear in mind:
  - There is no recognized international standard for monitors or foam application rates to be applied when using them for firefighting full surface tank fires with diameters over 20 m (65 ft). In fact, most standards suggest that monitors should not be used as the primary extinguishment method for such tank fires. Experience suggests that application onto the surface should be at least 8 l/min/m^2.
  - Calculations of the application rate and thereby the number of foam monitors required must account for foam losses due to foam stream drift, stream break-up, evaporation due to thermal effects, etc.
  - Foam stream ranges will always be listed in still air conditions but in reality will be affected by even a slight breeze.
  - Selecting low capacity monitors (5,000 l/min (1,320 g/min)), if they are able to reach the tank roof, will have an impact on the quantities of fire hose, foam pumpers, manpower and means of distributing foam concentrate.
  - Foam monitors selection should consider foam stream range, stability under working pressure and on rough terrain, portability/manpower required for deployment versus flowrate desired, remote and local foam concentrate pick-up capability and time taken to set-up for use.
- Foam pumpers should have either a balanced pressure proportioning system as described in NFPA 11, or similar method of foam pumping proportioning if they are to be used for foam supply to monitors. Pumpers which have round-the-pump-proportioning (RTPP) systems will not always be able to produce foam when working from a hydrant or other pressurized supply.
- Use of 200 litre (53 gallons) or 1,000+ litres (264+ gallons) containers as re-supply for foam pumpers must ensure that the pumpers have a means of picking-up (drafting) the concentrate from the containers as it will obviously be impossible to place these on vehicle roofs to drain into on-board tanks. The most efficient method of re-supplying foam pumper on-board tanks is to have the foam pump suction inlet valved to enable rapid changeover of foam containers as one is emptying. The best item of equipment for this is a collecting breeching (siamese) with 2 × valved inlets which connects onto the foam suction inlet and has connections compatible with the foam suction hose. Using this method and with large capacity containers of concentrate, only one pump operator per vehicle is required since one person can easily changeover containers by suction hose movement rather than having several personnel at each pumper.

- If hydraulic platforms, aerial ladders, fire vehicle roof mounted monitors or a combination of these are to be considered then the use of large capacity foam monitors (5,000 l/min+ (1,320 g/min)) on top of the ladders or platforms or fire vehicle roofs should be carefully examined to ensure the range will be suitable from the parking area. In addition, the number of such vehicles needed to create the required foam flowrate should be examined to ensure that they will normally be available and can access and park on tank area roads without blocking traffic for concentrate re-supply.
- Where foam tankers are to pump foam to the pumpers on-board foam tanks from topside, it should be noted that some agitation and therefore aeration is bound to occur and this may have an adverse effect on the foam concentrate supply.
- The knock-on effect regarding manning levels, hose requirements and overall water requirement of selecting portable/mobile equipment for full surface firefighting should be remembered and can be best illustrated by using an 80m diameter tank as the example of a full surface fire to be tackled.

> Total Surface Area = 5,027 m^2 (55,000 ft^2 rounded up)
>
> Application Rate = 10.4 $l/min/m^2$ (0.26 $g/min/ft^2$)
> (Considered rate accounting for foam stream losses)
>
> Total Application Rate = 52,281 l/min (14,000 g/min rounded up)
>
> Foam Concentrate = 3%
>
> Duration of Foam Application = 65 mins

(It is important to note that these 65 minutes should be complemented by other foam supplies to be sure to maintain the foam blanket after extinguishment.)

> Total Foam Concentrate = 52,281 × 0.03 × 65 = 101,948 litres (27,000 gallons)

For this example, and recalling that this application rate is considered the minimum, selecting 5,000 l/min (1,320 g/min) foam monitors would require 11 monitors and so would need slightly more foam concentrate since 11 × 5,000 = 55,000 l/min (14,520 g/min). Concentrate requirement would then be 55,000 × 3% × 65 mins = 107,250 litres (28,300 gallons). Also, 11 foam pumpers of at least 5,000 l/min (1,320 g/min) foam/water pump capacity would be needed. Assuming each foam pumper had 5,000 litre (1,320 gallons) foam tank on-board (55,000 litres (14,520 gallons) total) then several foam tankers or flatbed vehicles would be needed to distribute foam containers or foam concentrate direct to pumpers, possibly 132 × 75 mm (3 in) delivery hose for monitors, based on 12 for each monitor, 88 × 75 mm (3 in) soft suction hose from hydrants, based on 11 pumpers, and the manpower to deploy, monitor and reposition this equipment.

For the same example, selecting 15,000 l/min (4,000 g/min) foam monitors would require four and so would need more foam concentrate since 4 × 15,000 = 60,000 l/min (16,000 g/min). Consequently concentrate requirement would be 60,000 × 3% × 65 mins = 117,000 litres (31,000 gallons). Also, 12 foam pumpers or water pumpers of at least 5,000 l/min (1,320 g/min) water pump capacity would be needed, several foam tankers or flatbed vehicles to distribute foam containers or supply monitors directly, possibly 96 × 75 mm (3 in) soft suction delivery hose from hydrants, possibly 144 × 70 mm (3 in) delivery hose and 32 × 150 mm (6 in) delivery hose for monitors and the manpower to deploy, monitor and reposition this equipment.

If 30,000 l/min (8,000 g/min) foam monitors are selected, then two would obviously be required, or one at 60,000 l/min (16,000 g/min) capacity. The resources required for these would be similar to the 15,000 l/min (4,000 g/min) foam monitor example.

A point to note is that if reliance is placed on a single very large diameter foam monitor and this malfunctions during the fire, the firefighting efforts up to the point of failure will have been wasted.

Obviously, combinations of monitor capacity could possibly be used, depending on the effectiveness of the smaller capacity monitors when used alongside the larger monitors but the same logistics problems would exist and would need to be resolved before final selection of monitors.

- Fire hose sizes selection needs to consider the physical capabilities of firefighters to deploy them. The number of hoses required to be laid out may exceed one hundred and so the weight of hose becomes an important factor. Alloy couplings and a maximum diameter of 75 to 125 mm (3 to 5 in) and 20 metres (65 ft) length are typically used. Large diameter hose is increasingly used with very large capacity foam monitors. Again, alloy couplings will reduce weight but the length of these hoses should be limited to 3 or 4 metres (10 to 15 ft) for weight considerations unless mechanical deployment methods are possible. The maximum hydrant and pump working pressures should be checked to ensure that delivery hose will withstand the anticipated operating pressures. Particular attention should be paid to the strength and reliability of the large hose coupling attachment.

- Water supplies are a critical consideration for manual firefighting of a full surface tank fire. Using the previous example tank size of 80 m (260 ft), for the foam monitors water supply there may be water demands of up to 53,350 l/min (14,000 g/min) in the order of 10 bar (145 psi) (assume 11 × 5,000 l/min (1,320 g/min) monitors). Add to this possible water monitor cooling of radiant heat affected tankage or plant and it is obvious that total water demand could exceed 75,000 l/min (20,000 g/min). Water supply flowrate, pressure and availability must be carefully reviewed before considering manual firefighting, not only as a paper exercise but also as an actual test.

A7.10 Water run-off containment

Depending on the local bunding arrangement and the sewers/waste water treatment configuration, it may be necessary to provide a specific firefighting water containment area, such as the one illustrated in the picture below.

Alternatively, bunds not impacted by an emergency can be used to temporarily store contaminated water if adequate pumping capacity is available.

Water containment should be clearly addressed in the site Emergency Response Plan.

A7.11 Exercises

Frequent training and drills are also required to keep all staff involved up to date with their duties. Regular exercises of preplanned strategies are essential if an attack on a full surface fire is to be effective. This is particularly true when mutual help schemes are to be used in order to check compatibility of equipment, communications and understanding of each member's role and responsibilities. This was clearly demonstrated in the Sarnia naphtha tank fire in Canada where a mutual aid society acted in accordance with a well practised preplanned strategy and efficiently extinguished a 45m diameter fire.

Examples of rimseal fire training

The photographs above were taken during on-site tank workshops which combine classroom lectures with practical exercises. The workshops have been extremely useful in helping site personnel to understand their role in tank incidents. Such site specific workshops are strongly recommended.

The BP-run school held twice a year at College Station Texas A&M University provides for advanced exterior firefighting and command leadership training over five days/nights. The school is open to all BP&JV partners to attend and is run on a none profit basis with administration, instruction and fireground training carried out by BP fire chiefs and deputies. It is recommended for all sites to participate.

Appendix 8:
Some critical questions

1. Does your fire plan include scenarios and formal response strategies for:
   - rimseal fires;
   - full surface tank fires;
   - bund fires?
2. Are the water/foam needs calculated in accordance with this guidance recommended application rates?
3. Is there enough water pumping capacity to cover scenarios identified in 1?
4. Is there enough equipment and foam supply to cover scenarios identified in 1?
5. Is manpower adequate to cover scenarios identified in 1 and does everybody understand their role and responsibilities in any incident?
6. If answer is negative for 3, 4 or 5, what is the strategy in place (such as burn out, cooling) and is it written and formally agreed with relevant local authorities?
7. Is adequate training provided for large emergencies for:
   - operators;
   - plant firefighters;
   - external firefighters;
   - plant managers (for example, foam logistic, media training)?
8. Are large fire drills carried out, involving mutual help and local brigade equipment, with local authorities involvement?
9. Is the fire equipment in place regularly checked, tested and records kept (for example, foam quality and volumes, pumps tested, hydrants and pourers checked)?
10. Are back-up plans in place and tested in case of:
    - power failure;
    - fire main failure;
    - first foam attack failure;
    - protracted emergencies;
    - adverse weather during an emergency (such as snow, sand storm)?

Appendix 9: Learning from past accidents

Note: All drawings in this section were designed by René Dosne and are reprinted with permission.

A9.1 Spill and vapour explosion followed by fire at St. Herblain (France) depot

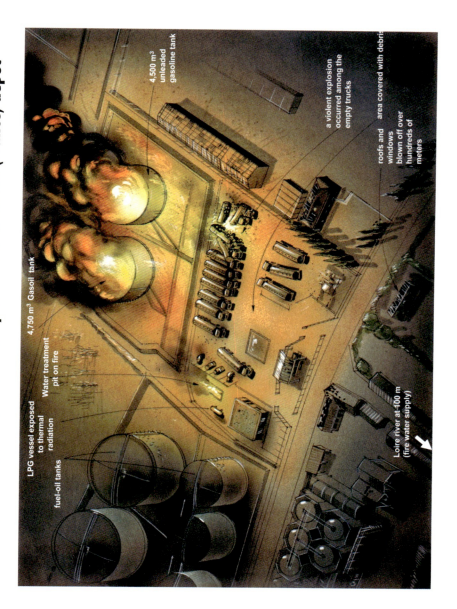

LIQUID HYDROCARBON STORAGE TANK FIRES

Circumstances

Early in the morning on 7th October 1991, a massive vapour cloud explosion occurred in a storage depot of approximately 80,000 m³ capacity. Near the two biggest tanks was a car park used to store distribution trucks overnight. At the time of the accident, around 4:20 am, the atmospheric conditions were as follows: 5°C, negligible wind speed, approximately 100% humidity.

The investigation team determined that a leak occurred on a transfer line when the depot valves were opened at 4:00 am for the first truck loadings of the day. Gasoline mixed with condensing humidity created a large visible cloud (around 23,000 m³, with a high around 1.5 m) which covered the storage area, a road and the parking. About 20 minutes later, the vapour cloud was ignited, probably by a truck engine, and a huge explosion followed. Two tanks (a 10,000 m³ tank containing 4,500 m³ of heating gasoil; and a 6,500 m³ tank containing 4,700 m³ of unleaded gasoline) of the depot caught fire.

The first response dealt with one fatality, three injured drivers and two injured operators, then focused on cooling two 15,000 m³ gasoil and gasoline exposed tanks and a small LPG vessel. Responders then focused on extinguishing fires outside the bund where the two tanks on fire were standing (these included truck fires, water treatment pit fire, etc.), while pumping resources were installed (the Loire river has a 8 m tidal difference, so pumps had to be located on a barge and a firefighting tug was also called) and foam collated. Once the pumping capacity was installed (approximately 30,000 l/min) and 80 m³ of foam available, the extinguishment sequence started with 15 foam monitors and was successfully completed around 12:17. More than 50 m³ of foam concentrate were used.

A study published in 1995 by the INERIS demonstrated that the intensity of the overpressure was mainly due to the confinement created by so many trucks parked closed together.

Lessons for tank fire responders

- Pumps have to be adequate for the local conditions. Large mobile pumps are now available with a downward reach of more than 50 m and a capacity above 24,000 l/min.
- Small monitors can be used successfully (as in that case) but they are manpower demanding and increase flexible hose deployment and congestion.
- Preplanning needs to include factors such as:
   - The major fire may not be immediately accessible due to explosion damage, debris, minor fires, etc.
   - Operators may be injured, leaving the incident commander with no local knowledge during the first phases of the response.
   - Local fixed systems (water pumps, monitors, foam stocks, etc) may need to be protected against blast/fire exposure.

A9.2 Tank fire at Abidjan, Ivory Coast depot, 13 May 1999

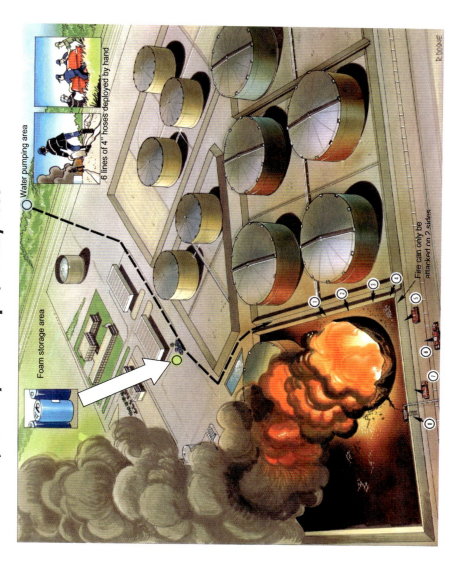

Circumstances

The GESTOCI terminal has 14 large storage tanks and is located near the Société Ivoirienne de Raffinage refinery. The fire at the GESTOCI terminal started around 14:45 on a Thursday on a 33,000 m³ gasoline tank B33 due to an unknown reason while the tank product was being transferred to a smaller tank. The fixed firefighting systems were insufficient and water truck shuttles also were needed to cool a nearby 20,000 m³ kerosene tank.

After two days, the Ivory Coast department requested international help and France dispatched 23 Paris fire brigade technicians, with foam and equipment. Helped by the military, they deployed water supply lines by hand from a water pumping area 600 m away (2,000 ft). Despite a first successful attack on Saturday, the fire restarted on Sunday and could not be reattacked before more foam stocks were flown into the country. The fire was finally extinguished on the following Tuesday afternoon.

Lessons for tank fire responders

- Fixed fire system (water supply, fire pumps, fire mains, foam stocks, etc) should be designed according to local risks and supplies.
- Emergency preplanning should identify required equipment and resources, and describe how to have them available on site quickly.

8 lines of 8" hoses layer truck

Typical 'hose spaghetti' issue

Laying of 6" hoses from containers

- Pumps have to be adequate for the local conditions. Large mobile pumps are now available with a downward reach of more than 50 m and a capacity above 24,000 l/min.
- Small hoses can be used successfully (as nearly achieved in that case), but they are manpower demanding and increase deployment time and congestion: Large hoses (6 to 12" (15–30 cm)) that can be deployed quickly by trucks are now commonly available (see pictures above).

LIQUID HYDROCARBON STORAGE TANK FIRES

A9.3 Spill fire at St. Ouen (France) depot, 14 June 1991: Hot work

LIQUID HYDROCARBON STORAGE TANK FIRES

Circumstances

Work was underway in the depot to convert some tanks to unleaded gasoline. The workbench used by the two workers was located on a path outside the tank farm, as was their diesel driven electricity generator. At 11:15, a small explosion occurred near the generator. The fire had soon covered 100 m², and was still spreading over the public railways towards a warehouse. The leak came from a drain line connected to a gasoline tank. Sand was used to create a bund to prevent the hydrocarbons from spreading further. For several hours, firefighters checked the valves at the bottom of the tanks without finding the one that was feeding the fire. While wading through the water-foam-hydrocarbon mixture, several firemen received burns due to the liquid under their feet catching fire. At 15:20, compressed air bottles on the site exploded, injuring eight responders. At around 16:00, the correct valve was closed and the fire stopped very quickly.

Fifteen people were injured including two seriously; three tanks were damaged; four railway lines and three nearby wagons were destroyed; 42 m³ of foam compound were used. The pumping of polluted firefighting water in the tanks lasted until 17 June.

Lessons for tank fire responders

- Large dimensional fires are impossible to put out as long as the fire is being fed (small or medium ones can be dealt with using dry powder or dual agents).
- Never commit personnel into a product spill, even covered with foam.
- Dual access should be available as far as possible to provide options for equipment deployment, in case of inaccessibility of one road and for escape.
- Preplanning needs to include factors such as:
  o Up to date drawings and installation knowledge should be available at all times.
  o External specialized resources may be needed 365 days per year, 24 hours per day, for urgent operations. These may include:
    – locomotives to move exposed railcars;
    – cranes to lift equipment in bunds or over damaged piping;
    – vacuum trucks and pumps to contain water/product run-off;
    – bulldozers to move sand, create bunds.
  o How to contact neighbours and what impact can they have on the depot activity and vice-versa.
  o Operators should be aware of hidden and transient risks (i.e. gas bottles for maintenance) and should communicate these to responders.

98

LIQUID HYDROCARBON STORAGE TANK FIRES

A9.4 Triple tank fire at Singapore refinery, 25 October 1988: Static spark from foam application

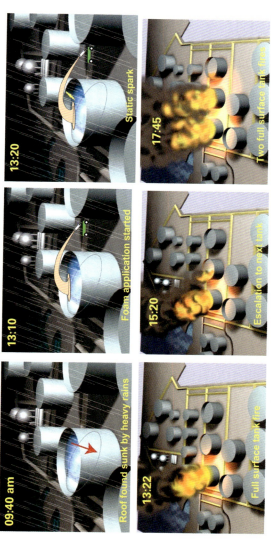

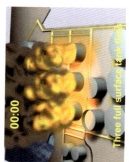

Circumstances

The largest petroleum fire experienced in Singapore occurred at 13:25 hours on the 25th October 1988 at SRC refinery. The incident involved three floating roof naphtha tanks containing a total of 294,500 barrels of product.

The three naphtha tanks shared a common bunded area. Each tank had a diameter of 41 metres (135 ft) with a shell height of 20 metres (66 ft). The tanks were spaced every 21 metres (70 ft).

In the morning, Tank 1116 TA was receiving sour straight run naphtha when the floating roof was found to be partially submerged. Filling of the tank was stopped and for two hours product was transferred out of the tank; however, this too was stopped when it was noted that the antirotational pole showing above the tank shell was physically displaced. The refinery fire service then began to apply foam but ignition occurred after ten minutes resulting in an immediate full surface fire. Two hours later, the fire escalated to adjacent Tank 1116 TB in the rim seal area and within a further two hours, it became totally involved. The third tank 1116 TC became fully involved at midnight the same day.

The fire was contained within the primary bund by cooling adjacent exposed tankage, and the naphtha tanks were allowed to burn out pumping out as much product as possible. The fire took five days to burn out.

The incident was dealt with by the combined efforts of the refinery personnel, the authorities and the mutual aid partners. No serious injuries occurred.

The enquiry team set up to investigate the incident concluded that the roof sank due to a combination of partially flooded roof pontoons and heavy local rainfall. Static created by foam application ignited the fire that it was supposed to prevent.

Lessons for tank fire responders

- Firefighters and operators should be aware of static generation hazards from water or foam application. It has become apparent that a number of tank fires which hitherto have been recorded as 'cause unknown' have been caused by static electricity generated during the application of foam from firemen's nozzles, remote monitors or even dripping from fixed foam pourers.
- Serious failures of floating roofs are not common occurrences, but they do occur frequently enough and with sufficiently serious potential consequences, to make it worth each site having a procedure in place for dealing with sunken or jammed roofs. Detailed actions will depend on the actual situation but in any case it is important that two steps are taken immediately:
  - Any movement of oil into or out of the tank should be stopped immediately.
  - The exposed surface of the oil should be covered in foam only when necessary and with precautions (see detailed section on this subject in this booklet).
- Pre-fire planning: Critical credible scenarios (fire on pumps, jetties or ships, in bunds or tanks, etc.) should be worked through in detail and drilled regularly. This will highlight any shortcomings in the available facilities, manpower or procedures. Testing the plan by means of exercises will identify any practical weaknesses. The mutual aid scheme and liaison between the site and the various authorities are also very important.
- Tankage layout: Escalation will occur if the layout is not adequate—see Appendix 3 of this booklet for more details.

A9.5 Fort Edouard Herriot depot, Lyon (France) – 2 June 1987

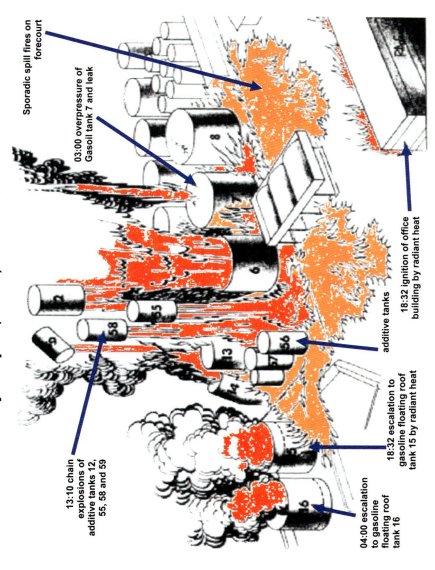

LIQUID HYDROCARBON STORAGE TANK FIRES

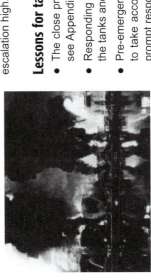

Circumstances

Two people were killed and 15 others injured in a fire at this oil depot. The site, which was built in the mid 1950s contained three tanks with gas oil and fuel oil, 40 tanks for other products and 50 small tanks containing additives. An additive tank was undergoing welding work. At 13:10, an explosion set fire to four additive tanks, and a fifth was blown into the air. While the plant firemen were unreeling the first hoses, a second explosion occurred, destroying their equipment. A third explosion blew an additive tank over 200 metres (650 feet) into the air. In view of the scale of the disaster, the authorities activated the Regional Emergency Plan (PPI) from 14:30 and evacuated the area. At around 17:30, a foam attack was tried on the 4,000 m² area of fire, after having gathered some 75 m³ of foam concentrate. At 18:32, a 1,000 m³ (6,000 bbl) tank of gas-oil experienced a boil-over (see right hand side picture) and collapsed. The wave of burning oil destroyed foam equipment and a fire truck and the fire escalated to nearby gasoline tanks. At 07:00 the next day, a new attack put out the bund fire and by 11:30, the fire was extinguished. Approximately 80% of the depot, including 14 tanks and a road tanker loading bay, and 10,000 m³ (>60,0000 bbl) product were destroyed.

It is suspected that the initiating fire was started by hot welding slag from welding work being carried out on the construction of a new tank.

Although the spacing between tanks within the depot was in accordance with local codes at the time of construction, it made the risk of escalation high.

Lessons for tank fire responders

- The close proximity of the tanks inevitably increased the risk of a fire escalating out of control—see Appendix 3 of this booklet for more details.
- Responding emergency services personnel need to have first hand information on the contents of the tanks and facilities available for emergency response on the site.
- Pre-emergency planning is an essential tool in reducing the impact of a major accident, and needs to take account of how industry and authorities work together to ensure that an effective and prompt response is made.
- Emergency equipment, including trucks, should not be located too close to the first fire—chances are that it will be lost in the case of escalation.

102

A9.6 Explosion of benzene tank 02 February 1989 in Villers St Paul (France): Welding work

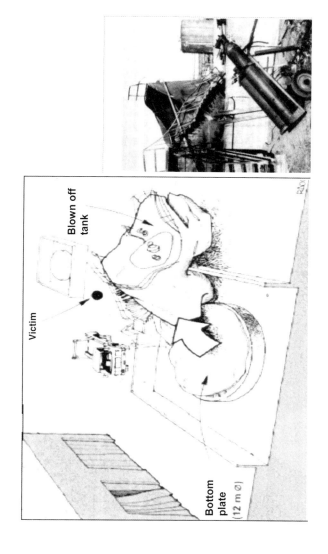

Circumstances

At about 9:40 am, three workers were performing work to upgrade fire-fighting piping on a benzene tank—this involved cutting old pipes and replacing them. After having worked with a pipe-cutting machine, an oxy-acetylene torch was brought to cut a flange on a pipe supplying the foam boxes. There was approximately 15 m^3 of benzene left in the tank.

An explosion took place, the tank was torn off and fell a few metres further in its bund, catching fire.

The worker who was using the torch died on the spot. His two colleagues were severely injured. Several persons were injured by glass debris and a sales person was hit by a door ripped off by the explosion's pressure wave.

The fire was extinguished in 40 minutes but the tank was destroyed.

It is likely that benzene vapours rose into the foam piping and the flame flashed back despite the presence of a (flame trap) screen in the foam box. The inquiry should establish whether this screen was worn out, if it had disappeared after a test of the foam boxes or if it had been occasionally removed from the tank.

Lessons

- A gas test must be carried out before any welding work takes place, even before starting work on a line which should only contain harmless substances.
- High-risk activities such as hot work and confirmed space entry require what is defined as positive isolation. This is achieved by inserting spades/blinds into lines or disconnecting lines to physically separate the work-piece from any possible introduction of hazardous materials. Refer to the BP booklet *Control of Work* in this series for more details.

Your notes